The Record of Native People on Gulf of California Islands

Thomas Bowen

Arizona State Museum
THE UNIVERSITY OF ARIZONA.

Arizona State Museum Archaeological Series 201

Arizona State Museum
The University of Arizona
Tucson, Arizona 85721-0026
(c) 2009 by the Arizona Board of Regents
All rights reserved.
Printed in the United States of America

ISBN (paper): 978-1-889747-83-5
Library of Congress Control Number: 2009930926
Second Printing, 2013

ARIZONA STATE MUSEUM ARCHAEOLOGICAL SERIES

General Editor: Richard C. Lange
Technical Editors: Lauren E. Jelinek and Rodrigo F. Renteria Valencia

The *Archaeological Series* of the Arizona State Museum, The University of Arizona, publishes the results of research in archaeology and related disciplines conducted in the Greater Southwest. Original, monograph-length manuscripts are considered for publication, provided they deal with appropriate subject matter. Information regarding procedures or manuscript submission and review is given under Research Publications on the Arizona State Museum website: *www.statemuseum.arizona.edu/research/pubs*. Information may be also obtained from the General Editor, *Archaeological Series*, Arizona State Museum, P.O. Box 210026, The University of Arizona, Tucson, Arizona, 85721-0026; Email: langer@email.arizona.edu. Electronic publications and previous volumes in the Arizona State Museum Library or available from the University of Arizona Press are listed on the website noted above.

The Arizona State Museum *Archaeological Series* is grateful to the many donors and supporters who continue to make this publication possible, particularly those who supported the 200th volume of the *Archaeological Series*.

Cover map by Tracy Davison.

Distributed by The University of Arizona Press, 355 S. Euclid Boulevard, Suite 103, Tucson, Arizona 85719

Contents

Figures

Figures, continued

Tables

Abstract

A century ago it was common knowledge among historians and anthropologists with research interests in northwestern Mexico that Indians had inhabited four large islands in the Gulf of California. Three of these islands—Cerralvo, Espíritu Santo, and San José—are in the far southern Gulf off the coast of Baja California. The fourth island, Isla Tiburón, is located on the mainland side more than 400 km to the northwest, in the constricted region of the Gulf known as the Midriff. At the turn of the twentieth century, Isla Tiburón was well known as the homeland of the Seri Indians and was the only Gulf island still inhabited by native people.

Since 1900, ethnohistorical and archaeological research has greatly expanded our knowledge of Indians on both sides of the Gulf. Much of that information, however, pertains to the people living on the peninsula and mainland, and touches on the islands only incidentally. Consequently, few historians or anthropologists are aware that Indians made use of many islands in addition to Cerralvo, Espíritu Santo, San José, and Tiburón. Scholars in other fields may not even know that there were Indians on the Gulf islands other than Tiburón. This is particularly unfortunate for ecologists because native people have been in the region for thousands of years and may have played a significant role in shaping the island ecosystems we see today. Although indigenous humans were by far the largest native terrestrial mammal on all but two islands, and the most voracious omnivore on all of them, *Homo sapiens* does not appear on island species lists, and the potential effects of native people on insular ecosystems have seldom been considered.

Discoveries made just in the past decade show that researchers in all fields have seriously underestimated the extent to which native people made use of the islands. Reports of early Spanish navigators have established the presence of Indians on many islands in addition to the four listed above. Recent archaeological research on several islands, along with fortuitous observations on others, have revealed evidence of native people on islands with no known documentary record of Indians. Chronological data from the southern Gulf establishes a time depth for indigenous people spanning at least ten millennia. New information from Seri oral history alludes to Seri voyages far beyond Isla Tiburón and greatly expands the picture of indigenous people in the Midriff region. Collectively, these results show that the traditional assumption, that most islands were beyond the range of native people, is dead wrong. It is now clear that Indians knew and exploited nearly every significant island in the Gulf.

This study reviews the evidence of native people on each of 32 major Gulf islands. The list includes all 22 islands larger than 2 km^2 and 10 smaller islands for which some data exist (summarized in Table 3.2). The data are drawn from historical documents, oral history, and the archaeological record. To the extent possible these data are given as quotations from the original sources.

Collectively, the evidence suggests that native people were familiar with at least 29 of the 32 islands. For 19 of the islands the evidence can be considered unequivocal, consisting of unambiguous historical documentation, credible oral history, and/or a clear archaeological record. For ten islands there is some evidence of human use, but it is limited, weak, or equivocal, and therefore, in need of corroboration. There are no data of any kind for two islands, and one small island has produced no evidence of native people despite a comprehensive archaeological survey (these data are summarized in Table 4.1).

Of course, Indians made greater use of some islands than others. In general, large islands, with a greater diversity of resources (including fresh water), were exploited more than small islands, and several supported permanent or seasonal communities of people. However, native visitors may have been drawn seasonally or intermittently to even very small islands with special resources, such as nesting seabirds, sea lion colonies, concentrations of cactus fruit, and abundant seed crops that appear after a rain. Historic and ethnographic sources show that islands did not need permanent water to sustain native visitors, who were quite capable of bringing water with them, subsisting on temporary water in tinajas, or utilizing water substitutes.

Distance was apparently no barrier to native use of islands. The cane *balsa* was the universal watercraft, and historic sources suggest that balsa traffic was extensive throughout the Gulf. In the hands of the Indians, the balsa was a swift and seaworthy craft, and navigation was no doubt facilitated by expert knowledge of winds and currents. All but two islands were within a day's paddle, and in most cases the overwater distance to the more remote islands could have been reduced by island hopping.

Although we now know that native people exploited nearly every significant island, we need much more information about the time span over which those visits took place. Indians were certainly making use of many islands during the seventeenth century when European navigators began keeping careful records. Seri oral history firmly places Seri ancestors on several Midriff Islands during the nineteenth century and conceivably earlier. Archaeology is potentially capable of extending island chronologies into the prehistoric past, but there has been only limited progress on this front. Fewer than half of the 32 islands considered in this study have been systematically surveyed, and controlled excavations have been conducted on only three islands. Unfortunately, the archaeological record for many islands consists only of surface remains, and includes few or no time-sensitive artifacts or structures, and little or no organic material suitable for radiocarbon analysis.

Clovis projectile points from both the Sonora mainland and the Baja California peninsula indicate that humans were present in the Gulf region by 13,000 years ago, which means that native people have been potential island visitors since that time. So far, radiocarbon dates have been secured for only five islands, and for four of these islands the few dates that are currently available all post-date AD 700. However, a series of 179 radiocarbon dates from 40 sites on Isla Espíritu Santo have clearly established this island's long occupational sequence, which extends from about 9000 BC to the fifteenth century AD. Although one site on this island produced spectacular radiocarbon ages on shells from approximately 36,550 to greater than 47,500 BP, these shells were probably already ancient when people collected them.

During the late Pleistocene and early Holocene it is likely that not all of today's islands existed, or existed as islands. Some small volcanic islands may not yet have emerged from the sea, while some of today's islands were connected to the shore by land bridges because sea level was much lower than it is today. Any island without a landbridge connection at that time would have been accessible only by watercraft. These non-landbridge islands should be considered prime places to search for early evidence of navigation.

This in turn raises the question of whether initial human entry into the Americas took place by boat along a Pacific coastal route, and whether subsequent dispersal involved the Gulf. In most coastal entry and dispersal scenarios, it is assumed that coastally-adapted boat people arriving at the southern tip of Baja California crossed the mouth of the Gulf to the Mexican mainland, making landfall in Sinaloa or even farther south. However, as R. James Hills has pointed out, this is highly unlikely. People arriving at the southern tip of the peninsula would have seen seemingly endless ocean in all directions except along the peninsular coast leading into the Gulf. There would have been no reason for people adapted to coastal resources to set out into an apparently empty sea. Instead, they would have followed the coastline into the Gulf, presumably exploring the islands along the way. The first reasonable place to cross the Gulf would have been the Midriff, where they could have island-hopped to the mainland with no overwater distance exceeding 17 km. In Hills' scenario, the Gulf would occupy a pivotal position in human dispersal in the Americas, and it is possible that evidence of this process has been preserved on some of the Gulf islands.

Resumen

Hace un siglo, era conocimiento común entre los historiadores y antropólogos con intereses de investigación en el noroeste de México, que algunos indígenas había habitado cuatro islas grandes en el Golfo de California. Tres de esas islas—Cerralvo, Espíritu Santo y San José—se encuentran en el lejano golfo al sur de la costa de Baja California. La cuarta isla, Tiburón, se localiza en el lado del continente a más de 400 km hacia el noroeste, en la región estrecha del golfo conocida como el Cinturón (*Midriff*). A principios del siglo XX, la Isla Tiburón era reconocida como el sitio de los indios Seri y era la única isla del golfo habitada todavía por gente nativa.

Desde 1900, la investigación etnohistórica y arqueológica ha incrementado enormemente nuestro conocimiento de los indígenas en ambas costas del golfo aún habitadas por gente nativa. Mucha de esa información, sin embargo, pertenece a la gente viviendo en la península y en el continente y se refiere a las islas solo incidentalmente. Consecuentemente, muy pocos historiadores o antropólogos están enterados que algunos indígenas utilizaron muchas otras islas además de Cerralvo, Espíritu Santo, San José y Tiburón. Es posible que estudiosos en otros campos ni siquiera sepan que hubo indígenas en las islas del golfo a no ser la isla Tiburón. Esto es particularmente desafortunado para los ecologistas porque la gente nativa ha estado en la región por miles de años y es posible que hayan jugado un papel importante en la configuración de las ecosistemas de las islas que vemos hoy en día. Aunque los indígenas humanos eran seguramente los mamíferos terrestres nativos más grandes en todas las islas, a excepción de dos islas, y el omnívoro más voraz de todos ellos, *Homo sapiens* no aparece en las listas de especies de la isla, y los efectos potenciales de la gente nativa en los ecosistemas insulares raras veces han sido considerados.

Los descubrimientos recientes hechos en la década pasada muestran que los investigadores en todos los campos han subestimado seriamente la intensidad en que la gente nativa ha utilizado las islas. Los reportes de los primeros navegantes españoles han establecido la presencia de indios en muchas islas además de las cuatro arriba mencionadas. La investigación arqueológica reciente en varias islas, así como investigaciones fortuitas en otras islas, ha revelado evidencia de gente nativa en las islas, si bien no hay récordes conocidos que documenten la presencia de los indígenas. Los datos cronológicos del golfo sureño establecen un período de tiempo para la gente indígena que se extiende por al menos diez milenios. Nueva información de historia oral de los Seris alude a viajes de los Seris más allá de la Isla Tiburón y amplía grandemente la imagen de la gente indígena en la región del Cinturón. En conjunto, estos resultados muestran que la suposición tradicional de que la mayoría de las islas se encontraban más allá del rango de las gentes nativas, es totalmente errónea. Es claro ahora que los indígenas conocieron y explotaron prácticamente cada isla significativa en el golfo.

Este estudio revisa la evidencia de gente nativa en cada una de 32 islas importantes del golfo. La lista incluye todas las 22 islas mayores de 2 km² y 10 islas más pequeñas de las cuales existen información (ver Tabla 3.2). Los datos se obtienen de documentos históricos, historia oral, y récordes arqueológicos. En la medida de lo posible, esta información se presenta citando fuentes originales.

En su conjunto, la evidencia surgiere que la gente nativa estaba familiarizada con al menos 29 de las 32 islas. En 19 de esas islas, la evidencia se puede considerar inequívoca, consiste en documentos históricos sin ambigüedad, historia oral creíble, y/o récordes arqueológicos concisos. Para 10 islas hay alguna evidencia de uso humano, pero es limitado, débil o equivoco y por lo tanto, necesita corroboración. No existen datos de ninguna clase para dos de las islas, y una isla pequeña no ha producido ninguna evidencia de gente nativa a pesar de una investigación arqueológica exhaustiva (ver Tabla 4.1).

Evidentemente, los indígenas utilizaron unas islas más que otras. En general las islas grandes, con mayor diversidad de recursos (incluyendo agua dulce), fueron explotadas más que las islas pequeñas, y muchas de ellas apoyaron comunidades de personas permanentemente o por temporadas. Sin embargo, los visitantes nativos deben haber sido atraídos por temporada o intermitentemente hacía las islas muy pequeñas con recursos especiales, tales como aves marinas anidando allí, colonias de lobos marinos, concentraciones de frutos de cactáceas, y cosechas abundantes de semillas que aparecen después de la lluvia. Fuentes históricas y etnográficas muestran que las islas no necesitaban tener agua permanentemente para sostener la vida de los visitantes, quienes eran bastante capaces de traer agua con ellos, subsistir en agua acumulada temporalmente en tinajas o de utilizar sustitutos de agua.

Aparentemente, la distancia no era una barrera para el uso de las islas. Las balsas de carrizo eran el medio de navegación universal, y fuentes históricas surgieren que el tráfico en balsa era ampliamente utilizado en todo el golfo. En las manos de los indígenas, la balsa era una embarcación rápida y eficiente en el mar, y la navegación indudablemente era facilitada por el conocimiento experto de los vientos y las corrientes. A excepción de dos islas, todas las demás estaban a una distancia de un día de remo, en la mayoría de los casos la distancia en mar abierto de las islas más remotas se podía reducir pasando de isla a isla.

Aunque sabemos ahora que la gente nativa explotó prácticamente cada isla significativa en el golfo, necesitamos mucha más información sobre el período de tiempo en el que se llevaron a cabo dichas visitas. Ciertamente los indígenas estaban utilizando muchas islas durante el siglo XVII cuando los navegantes empezaron a llevar récordes detallados. La historia oral de los Seri ubica muy contundentemente a los ancestros Seri en varias islas del Cinturón durante el siglo XIX y posiblemente antes de eso. La arqueología es potencialmente capaz de extender la cronología de las islas hasta el pasado prehistórico, pero el progreso en este campo ha sido limitado. Menos de la mitad de las 32 islas consideradas en este estudio han sido investigadas, y sólo en tres islas, se han llevado a cabo excavaciones controladas. Desafortunadamente los datos arqueológicos para muchas islas consisten solamente de vestigios superficiales, e incluyen pocos artefactos o estructuras que no tienen relevancia significativa de tiempo, y hay poco o ningún material orgánico que sea válido para análisis de radiocarbono.

Las puntas Clovis provenientes de la parte continental de Sonora y la península de Baja California indican que los humanos estaban presentes en la región del golfo hace 13,000 años, lo que significa que la gente nativa ha sido visitante potencial de las islas desde entonces. Hasta ahora, se han asegurado fechas de radiocarbono para cinco islas solamente, y para cuatro de esas islas las pocas fechas accesibles hasta hoy, todas tienen una fecha posterior a 700 años d.C. Sin embargo, una serie de 179 fechas de radiocarbono de 40 sitios en la isla Espíritu Santo han establecido claramente la secuencia de ocupación de esta isla que se extiende desde aproximadamente 9,000 a.C. hasta el siglo XV d.C. A pesar de que un sitio en esta isla produjo edades espectaculares de radiocarbono en conchas de aproximadamente 36,550 a más de 47,500 años, estas conchas probablemente ya eran arcaicas cuando la gente las recogió.

Es posible pensar que durante el Pleistoceno tardío y el Holoceno temprano no existían todas las islas que en la actualidad se hallan presentes en el golfo de California, o que al menos no existían como islas. Es posible que algunas islas pequeñas no hubieran emergido todavía del mar, en tanto que algunas de las islas actuales estaban conectadas a la costa a través de puentes terrestres porque el nivel del mar era más bajo de lo que es hoy en día. Cualquier isla sin una conexión de puente terrestre en ese tiempo, habría sido accesible únicamente a través de una embarcación. Estas islas sin puente terrestre deberían considerarse como los lugares óptimos para buscar evidencias tempranas de navegación.

Lo anterior permite cuestionar si la entrada inicial de humanos en las Américas tuvo lugar a través de una ruta a lo largo de la costa Pacífica mediante el uso de embarcaciones, o si subsecuentes dispersiones migratorias involucraron al golfo. En la mayoría de los escenarios de entrada por la costa o por dispersión, es asumible que las personas navegantes adaptadas a la costa, llegaron a la punta sur de Baja California y cruzaron la boca del golfo hacia al continente mexicano llegando a Sinaloa o incluso más lejos hacia el sur. Sin embargo, como ha señalado R. James Hills, esto es altamente improbable. La gente que llegó a la punta sur de la península debe haber visto lo que aparecía como un océano sin fin en todas direcciones excepto a través de la costa peninsular que llevaba hasta el golfo. No habría habido razón para la gente adaptada a los recursos de la costa para embarcarse hacia un aparentemente mar vacío. En lugar de esto, ellos habrían seguido la línea costera hacia el golfo, posiblemente explorando las islas en el camino. El primer lugar razonable para cruzar el golfo habría sido el Cinturón, donde ellos podrían haber brincado de isla a isla hasta el continente, con una distancia sobre el mar de menos de 17 km. En el escenario de Hills, el golfo ocupa una posición fundamental en la dispersión humana en las Américas y es posible que la evidencia de este proceso haya sido preservada en algunas de las islas del golfo.

Acknowledgments

This publication came about because of an informal chat with Gary Nabhan ten years ago. It was a conversation about the possible role of native people in shaping modern ecosystems on Gulf of California islands. As Gary pointed out, ecologists working in the Gulf are all too aware of changes wrought by modern humans, but few realize that indigenous people may have altered island ecosystems over hundreds, if not thousands, of years. Since that conversation we have both explored this topic, he directly and extensively (Nabhan 2000, 2002, 2003); I briefly and tangentially (Bowen 2003, 2004). However, these discussions focused on the Seri Indians and the few Midriff islands we knew they had visited. Neither of us knew much about the indigenous history of Gulf islands as a whole, and a little searching showed that nobody else did either. It was obvious that ecologists, biogeographers, or any one else who wished to understand the role of native people on the islands would need better and more systematic information than was readily available, and that it would be helpful to have it all in one place. This publication is the result. Thanks, Gary, for planting the idea.

Many people have contributed to this project by generously sharing their knowledge of the islands. Historian Mike Mathes was extremely helpful in steering me to key documents pertaining to the early European exploration of the Gulf. Linguists Steve Marlett and Cathy Moser Marlett were instrumental in sorting out Seri names of the Midriff Islands and for drawing my attention to the *Hant Ihiini* people. Ricardo Arce, Peter Butz, Richard Felger, Doug Peacock, Robert Rossiter, José Smith, Charlie Sylber, Bernie Tershy, Rito Vale, and Ben Wilder filled me in on the availability of water on several islands. Conrad Bahre, Elfego Briseño, Lloyd Findley, Harumi Fujita, Peter Garcia, Gordon Gastil, Luke George, Jim Hills, Eric Ritter, Steve Shackley, Judith Thatcher, Nancy Trippe, Enriqueta Velarde, Elisa Villalpando, John Weed, and Rich White all contributed unpublished information on the archaeology of the islands. I thank Jim Hills for allowing me to present his hypothesis of initial human entry into the Gulf. Additionally, I owe a truly enormous debt of gratitude to linguists Ed and Becky Moser who unstintingly shared their vast knowledge of the Seris with me over not just years, but decades. I am deeply grateful to you all.

Much of the information presented here is based on archaeological research, and many people have played important roles in making my own fieldwork possible. I am indebted to the Instituto Nacional de Antropología e Historia in Mexico City, and its regional offices in Sonora and Baja California, for permits and permission to conduct fieldwork on the islands. I particularly thank Beatríz Braniff, Francisco Mendoza, Xicoténcatl Murrieta, Arturo Oliveros, Cynthia Radding, and Elisa Villalpando of the Sonora office, and Julia Bendímez of the Baja California office, for their extensive help in expediting the permitting process. I am also indebted to Roberto Herrera of the Seri tribal council, and to Armando Araujo of the Sonora office of Fauna Silvestre for arranging permissions from their agencies to conduct field work on Isla Tiburón.

Nearly all the fieldwork on the Midriff Islands since the mid-1970s has been observational—conducted with no collecting or excavation and hence no disturbance to the archaeological record. Since 2004, this work has been carried out as an "Inventory of Cultural Resources" under the auspices of Área de Protección de Flora y Fauna Islas del Golfo de California. I am grateful to Alfredo Zavala and Carlos Godínez of the Ensenada office of this agency, to Erick González, David Ramírez, Isabel Arce, and Hugo Moreno of the Islas office in Bahía de los Ángeles, to Ana Luisa Figuroa of the Guaymas office, and to Karina Santos of the parent agency, the Instituto Nacional de Ecología, for their unflagging support of this project and their continuing friendship.

Fieldwork on islands can be a difficult undertaking, and I have had invaluable logistical help from Samuel Díaz, Carolina Espinoza, Eldon Heaston, Joan Heaston Jarratt, Larry Johnson, Tad Pfister, Tish Rankin, and Wes Rankin. Three superb backcountry pilots—Sandy Lanham, Pat Patterson, and Ike Russell—helped immeasurably in preparing for field trips by providing low-level reconnaissance flights over the Midriff islands. But conducting field work on islands inevitably requires boats. Crossing the sometimes treacherous Gulf can be a real adventure, and I am deeply grateful for the superb skill and judgment of the Mexican and Seri panga fishermen and North American sport sailors who have taken us to the islands. Over many years, these mariners have included Dan Anderson, José Arce, Ricardo Arce, Chapo Barnet, Guillermo Cardenas, Martín Cortés, Feliciano Cota, Raul Espinoza, Alan Ferraris, Igor Galván, Emilio García, Eldon Heaston, Bob Jarratt, Larry Johnson, Severiano León, Alberto Lucero, Hugo Moreno, Tad Pfister, José Smith, George Weary, and Bill Zuliger. I especially thank Larry Johnson for all the times he delivered us safely in seas that would make lesser sailors cringe, and for saving our bacon when it looked as if we would lose the boat.

Much of the joy of field work lies in the competence and character of those with whom you share the experience, and over the years I have been fortunate to work with talented and dedicated people who are both colleagues and friends: Jon Avent, Carol Avent, Dan Bench, Dave Bockoven, Diane Boyer, Marty Brace, Bill Broyles, Dana Desonie, Jonathan Hanson, Roseann Hanson, Steve Hayden, Jim Hills, Larry Johnson, Elisa Villalpando, and Rich White. Many of us, in fact, have joined forces repeatedly, and in this context I wish single out two individuals. Steve Hayden and I have a field partnership that spans more than 40 years, from our first island trip in July 1967 (when we learned a thing or two about summer heat and insufficient water), to our most recent field trip in January 2009. Dan Bench and I have collaborated on 13 trips between 1977 and 2006, totaling almost 150 days in the field together. Thank you Dan and Steve, and my thanks to all of you for your tireless efforts, astute observations, and your wonderful companionship.

Many people have also contributed to the preparation of the manuscript. I stand in awe of the creativity and perseverance of inter-library loan librarians Ann Harris, Maggie Kanengieter, and Barb Oakleaf in tracking down obscure historic sources, and I thank Tom Duncan for the care with which he checked my translations of several of those documents. I am indebted to Tracy Davison for her generous investment of time and computer expertise that converted my crude scribblings into elegant maps. Harumi Fujita kindly supplied photos of Islas Cerralvo and Espíritu Santo and gave permission to publish them. The National Museum of the American Indian and the Department of Special Collections at the University of Arizona Library granted permission to publish Figures 2.6 and 3.10 respectively. Scott Copeland prepared the photos for publication, and I thank him for his artistry and technical wizardry. The abstract was translated to Spanish by Rodrigo Valencia and Tom Duncan, whom I thank for their effort and care in this task. Marty Brace, Loren Davis, Richard Felger, Alan Ferg, Harumi Fujita, Jim Hills, Mike Mathes, Don Laylander, and an anonymous reviewer read the manuscript in part or in full, and I am grateful to them all for their thoughtful comments and suggestions. To Rich Lange, editor of the ASM Archaeological Series, I offer my sincere appreciation for rapidly and expertly shepherding the manuscript from initial inquiry to publication with a collegial friendliness that is all too rare these days.

And finally, I offer my heartfelt gratitude to Marty Brace for moral support, sage advice, and most of all for cheerfully putting up with all those long absences and seemingly endless hours at the keyboard. Mil gracias, Marty!

Chapter 1
Background

A century ago, it was common knowledge among historians and anthropologists with research interests in northwestern Mexico that Indians had inhabited four of the large islands in the Gulf of California (Fig. 1.1). Three of these islands—Cerralvo, Espíritu Santo, and San José—are in the far southern Gulf off the coast of the Baja California peninsula (Fig. 1.2). The historic record of Indians on these three islands begins in the late sixteenth century when Europeans sailed into the southern Gulf seeking personal wealth in pearls. The record ends in the 1730s when the Jesuit missionaries who colonized the peninsula persuaded the Isleños (as they were often called) to abandon their island homes and relocate at the peninsular missions. By the late eighteenth century most of the Indians themselves were gone, victims of European diseases and acculturative pressures. But some of the Europeans who knew the Isleños at first hand wrote of their experiences with them. From the letters, reports, bureaucratic paperwork, memoirs, and histories available to scholars a hundred years ago, it was clear that Islas Cerralvo, Espíritu Santo, and San José had been inhabited by native people.

Anthropologists and historians of the period were also familiar with the first archaeological investigations on the southern islands. In 1883 the Dutch anthropologist Herman ten Kate recovered several burials from Isla Espíritu Santo (Kate 1977:57-61), and between the 1890s and early 1900s the French naturalist Léon Diguet excavated burial caves on both Espíritu Santo and Cerralvo (Diguet 1898:41-44, 1973:27). The human remains they recovered, secondary bundle burials of ochre-painted bones, were of considerable academic interest, in no small measure because the early European observers had made no mention of this peculiar form of interment among any of the peninsular peoples.

The fourth island that everyone knew was inhabited by Indians was Isla Tiburón. This island, situated on the mainland side of the Gulf, lies more than 400 km northwest of the three southern islands, in the constricted region of the Gulf known as the Midriff (Fig. 1.3). By 1900, a profusion of historical documents spanning nearly three centuries, plus the then-recent ethnographic and archaeological fieldwork of WJ McGee (1898), had firmly established Isla Tiburón as the traditional homeland of the Seri Indians (or Comcáac, as they refer to themselves). At the dawn of the twentieth century, it was the only Gulf island inhabited by indigenous people, and Seri activities on Isla Tiburón were still sensational news in both Mexico and the United States (Bowen 2000:243-258).

Since 1900 research in several field has greatly expanded our knowledge of both the peninsular Indians and the mainland Seris. Historians in Mexico, Spain, and the United

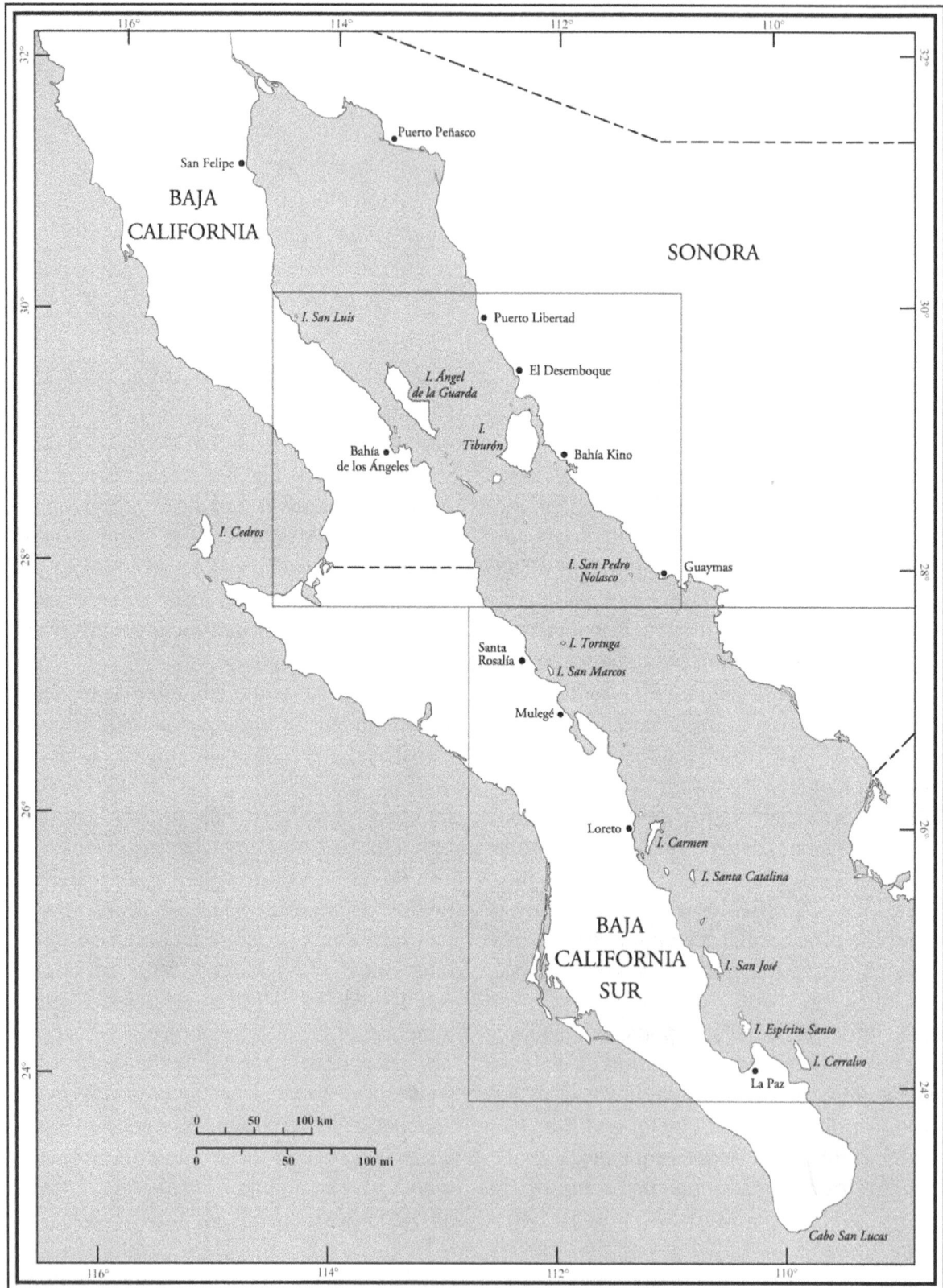

Figure 1.1. The Gulf of California. Lower and upper boxes indicate areas covered by Figure 1.2 and Figure 1.3 respectively.

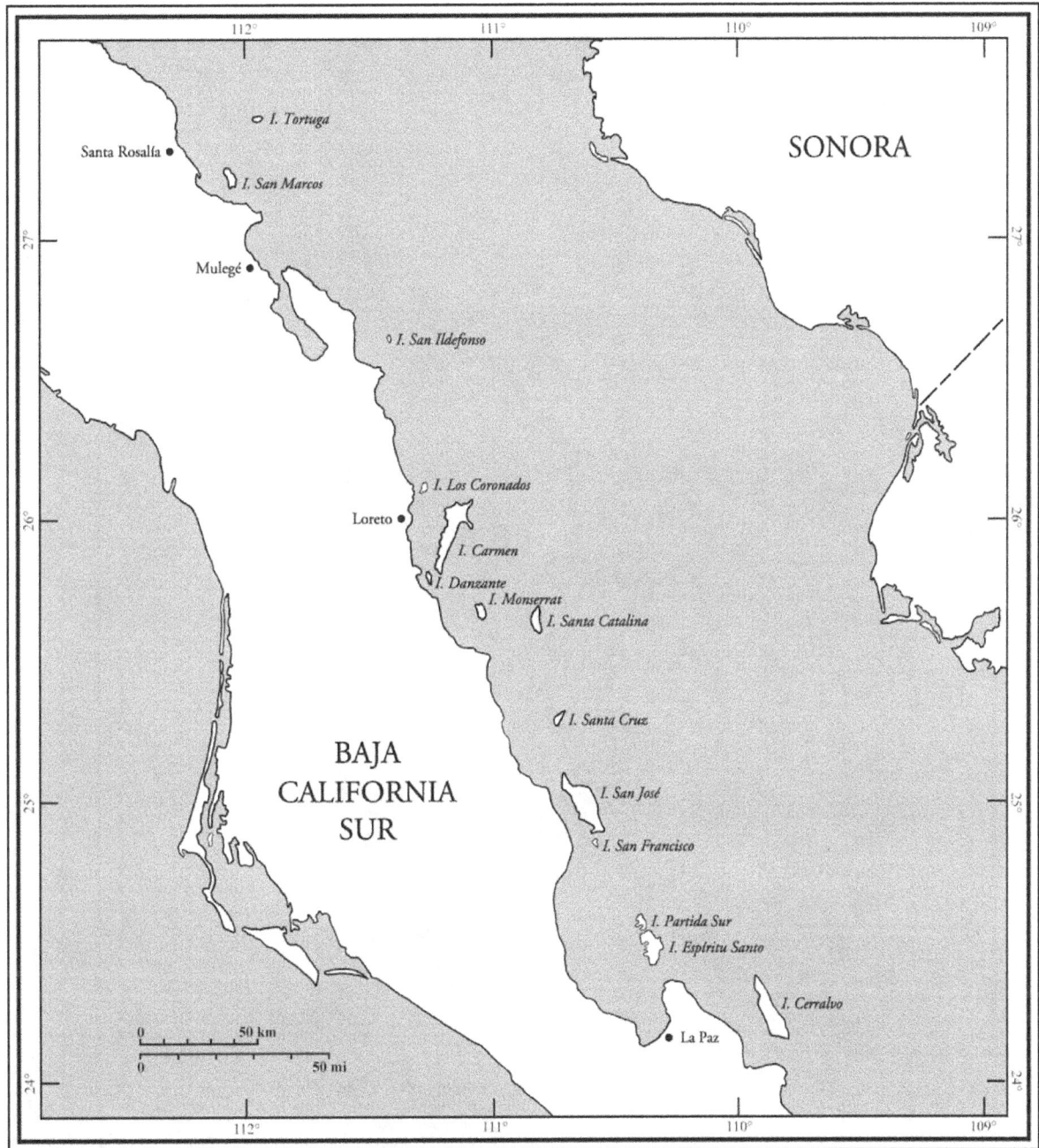

Figure 1.2. Islands of the southern Gulf.

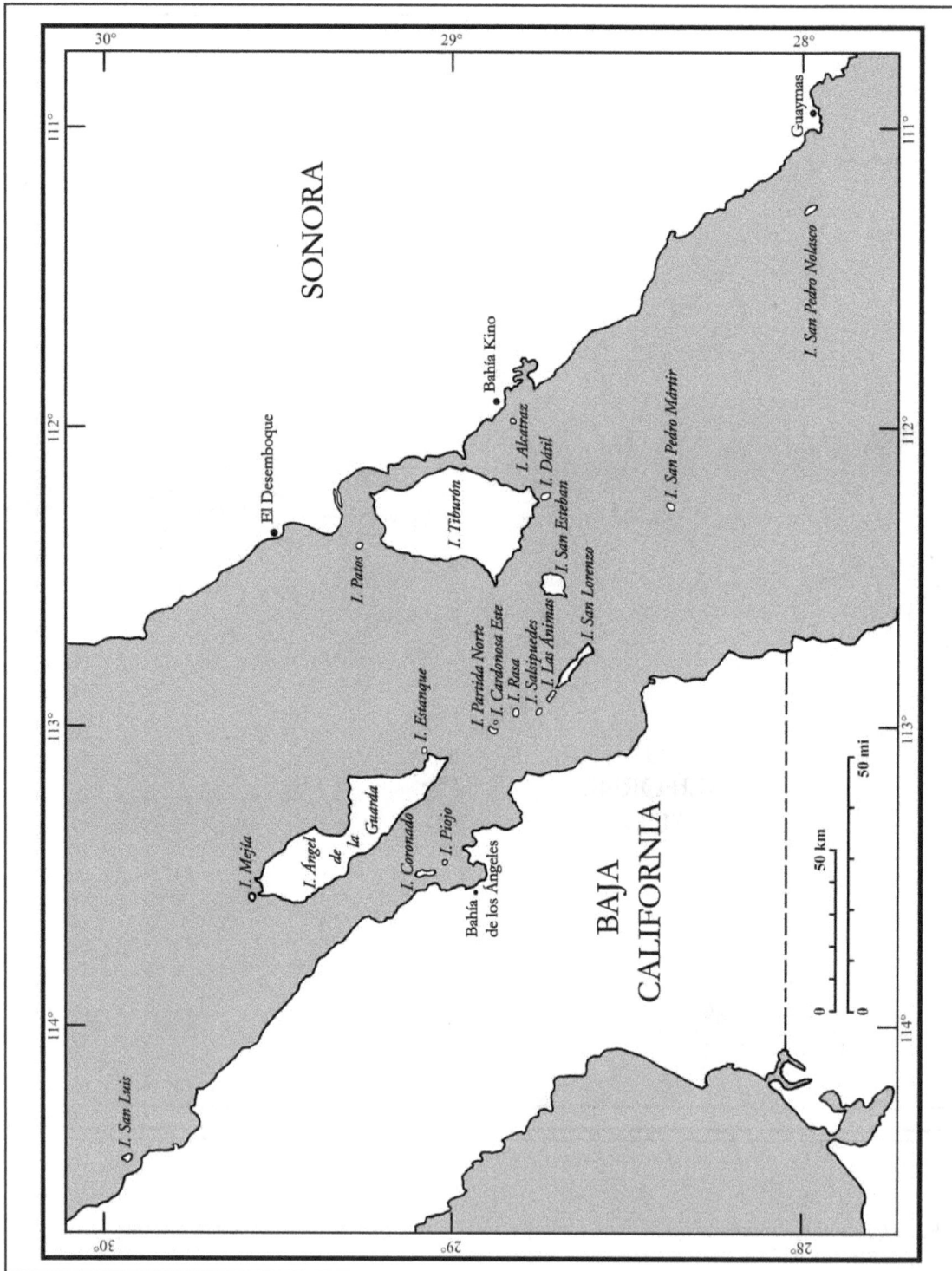

Figure 1.3. Islands of the Midriff region.

States have performed an enormous service by locating, assembling, organizing, indexing, copying, transcribing, and publishing a broad range of original documents unknown or not readily available to nineteenth century scholars. For Baja California, this has greatly facilitated efforts to reconstruct the ethnographic and linguistic history of the peninsular peoples (for example, Aschmann 1959; Massey 1949; Mathes 2006; Mixco 2006). Meanwhile, archaeologists and rock art specialists have explored much of the length and breadth of the peninsula and extended the cultural record of its aboriginal inhabitants back to the Pleistocene-Holocene transition (Laylander 1992; Reygadas 2003), paving the way for the first modern summary of peninsular prehistory (Laylander and Moore 2006). In recent years, anthropological and ethnohistorical research in Baja California has joined the internet revolution with projects such as Don Laylander's exhaustive on-line bibliography and compendium of peninsular radiocarbon dates (Laylander 2004, 2007).

For the Sonora mainland, documents compiled over the past hundred years have been equally important in enhancing our understanding of Seri history (for example, Sheridan 1999). Unlike the peninsular groups, however, Seri culture and language are known in great detail from consultation with living individuals. McGee's fleeting stint among the Seris aside, substantive ethnographic and linguistic field work was initiated in 1930 by Alfred Kroeber (1931), extended in mid-century by William Griffen (1959) and Margarita Nolasco (1967), and supplemented with descriptions of specific aspects of Seri culture by a number of observers (for example, Cano ca. 1960; Hinton 1955; Ryerson 1976; Smith 1974; Woodward 1966; Xavier 1946; see also Bowen 1983). But Seri ethnography, linguistics, and oral history as we know them today are largely the legacy of Edward Moser and Mary Beck Moser, who lived and worked with the Seris for much of the second half of the twentieth century. They and their collaborators and colleagues have produced a rich and nuanced portrait of these people, their culture, and their language (for example, E. Moser 1963, 1973; M. Moser 1970, 1978, 1988; Moser and Moser 1961, 1965, 1976; Moser and White 1968; Moser and Marlett 1998, 2005; Bowen and Moser 1968; Felger and Moser 1985; Marlett 1984, 1990; Bowen 2000:5-30; see also Wilder 2000).

On the other hand, investigations of the archaeology of the Seri region have lagged far behind both Seri ethnology and peninsular anthropology. Published information as late as the 1960s consisted mostly of reports of single artifacts and small collections (for example, Dockstader 1961; Holzkamper 1956; Moser and White 1968) or brief field observations (for example, Fay 1961; Hayden 1956). Although limited surveys and small-scale excavations have been conducted in recent decades, most remain unpublished and thus not widely known (for example, Bowen 1986; Villalobos 2007; White 1975). Sadly, accessible information on the Seri region still consists mainly of the published accounts of two aged archaeological surveys, one from the 1960s (Bowen 1976) and the other from the 1980s (Bowen 2000:315-394; Villalpando 1989), along with reports of a few sites at the outer margins of Seri territory (Bowen 1976:110-115, 2005a; Dixon 1990).

Much of the information accumulated during the past century, both historical and archaeological, concerns Baja California and mainland Sonora per se, and touches on the islands only incidentally. Consequently, while historians familiar with the reports of the early Spanish navigators have long known that Indians made use of islands other than Cerralvo, Espíritu Santo, San José, and Tiburón, most anthropologists still do not. Scholars in other fields may not even be aware that there were Indians on Gulf islands other than Tiburón (Case and Cody 1983:vii, 1987:408). This is particularly unfortunate for ecologists because native people have been in the Gulf region for some 13,000 years and may have played a significant role in shaping the island ecosystems we see today. Although indigenous humans were by far the largest native terrestrial mammal on all but two islands, and the most voracious omnivore on all of them, *Homo sapiens* does not appear on island species lists (for example, Carabias and others 2000:Anexo 3; Case and others 2002:Appendix 12.1; López-Forment and others 1996), and their potential effects on insular ecosystems have seldom been considered (Nabhan 2000, 2002, 2003; for a list of important exceptions, see Bowen 2004:200).

Discoveries made just in the past decade show that we have all seriously underestimated the extent to which native people made use of the Gulf islands. New information from Seri oral history alludes to Seri voyages far beyond Isla Tiburón and greatly expands the picture of indigenous people in the Midriff region. Recent archaeological research on islands in both the southern Gulf and the Midriff, along with fortuitous observations on other islands, provide evidence of native people on islands with no known documentary record of Indians. Chronological data from the southern Gulf establishes a time depth for indigenous people spanning at least ten millennia. Collectively, these results show that the traditional assumption, that most islands were beyond the range of native people, is dead wrong. It is now clear that Indians knew and exploited nearly every significant island in the Gulf.

The purpose of this publication is to review current knowledge of native people on Gulf of California islands. Chapter 2 provides an overview of the data in order to establish their relevance and address matters of reliability and limitations. Chapter 3 presents evidence of native people for each island in turn, drawing on historical documents, oral history, and the archaeological record. To the extent possible, data are given as direct quotations from original sources. The final chapter summarizes the evidence and considers briefly some of its implications.

Chapter 2
Overview of the Data

The Documentary Record

Some of the first Europeans to explore the Gulf of California left detailed descriptions of Indians, or detailed descriptions of islands, but not much about Indians on islands. Francisco de Ulloa, who circumnavigated the Gulf in 1539, carefully described many of the islands in terms that make them identifiable today. However, there is no indication that he went ashore on any of them, so there is little reason to trust his opinion of them as "uninhabited" (Ulloa 1924:322, 329-330, 332). In 1596, Sebastián Vizcaíno explored the southern Gulf and landed on several islands. Although he found Indians on some, most islands are not identifiable, and his report is exuberantly contradictory:

> I found a great number of uninhabited islands (Vizcaíno 1930:210).

> The islands here...were full of warlike Indians (Vizcaíno 1930:215).

> [There was] no island among the many I discovered where there were no inhabitants (Vizcaíno 1930:217).

Among the early reports, the terms "uninhabited" and "uninhabitable" often embody the biases of their writers and must be interpreted cautiously (Aschmann 1965). In some cases these terms may mean incapable of supporting European settlers dependent on agriculture. In others, they may mean not *permanently* inhabited. For example, Gonzalo de Francia, who accompanied Vizcaíno in 1596, wrote of Isla Espíritu Santo:

> It is uninhabited as people only go there in summer in some small cane *balsas* (Francia 1930:219).

Moreover, early European visitors rarely went ashore at more than one or two locations and they seldom stayed on islands very long. Those who failed to find Indians may have landed at places Indians did not frequent or arrived at the wrong time of year. In some cases, Indians may have hidden themselves, as they sometimes did after brutal encounters with unscrupulous pearl hunters. For the southern Gulf, the most important early source of information is Francisco de Ortega, who explored the islands for their pearl resources during three separate voyages in 1632, 1633, and 1636 (notwithstanding historian Ernest Burrus's assertion [1972] that Ortega fabricated the 1636 voyage; see Bowen [2000:479-481] and Mathes [1991]). Ortega described most of the islands in identifiable terms and gave many their present names. His meticulous record of where he found Indians and their shell middens often provide the earliest reliable documentation.

After Ortega, documentary information

for the southern islands comes mainly from the Jesuit missionaries who labored on the peninsula from 1697 to 1768. These men were prolific writers, and islanders figure in many of their letters and reports, occasionally in some detail. Although in rare instances Jesuits visited the Isleños on their home islands (for example, Bravo 1970), their policy was to relocate them more conveniently to the peninsular missions. By the 1730s the islands had been depopulated and thereafter served mainly as places of refuge for rebel Indians on the run, effectively ending native occupation. Thus while only the early missionaries knew the Isleños when they were still island people, the later Jesuits all knew of them and sometimes mentioned them in their memoirs (for example, Baegert 1952:12; Barco 1981:20). Even eighteenth-century Jesuit historians who resided comfortably in the heart of New Spain and Europe, and relied for their data on reports from their brethren in the field, knew about the Isleños (for example, Clavigero 1937:86-87, 255, 259; Venegas 1979:374).

In the Midriff region, Seri island history begins with a pearl hunting expedition led by Nicolás de Cardona and captain Juan de Iturbe, who went ashore on Isla Tiburón in 1615 (Cardona 1974:102-104). By 1645, Jesuit missionaries in the interior of Sonora were aware of Seris living on Tiburón (Pérez de Ribas 1999:23). Since that time there has been a nearly continuous record of Seris on that island, virtually to the present (Bowen 2000).

Oral History

History, for non-literate peoples, is a matter of passing along memories of past events through the generations by word of mouth, and some cultures have been adept at keeping these memories alive over long spans of time. However, if the people die out, their history dies with them unless their traditions have been preserved in writing by outsiders. Unfortunately, whatever oral history the Baja California Indians may have had, little was recorded by the Europeans, and apparently nothing at all concerning the islands.

For the Seris, it is an entirely different story. The Seris have an astonishingly rich corpus of oral history. Efforts to record it in depth, both in writing and on tape, began in the 1950s with the pioneering work of Edward and Mary Beck Moser, and it continues today, increasingly by Seris themselves (Herrera 2009; Stephen Marlett, personal communication 2007). To appreciate the extent and depth of Seri oral history, it is well worth perusing Felger and Moser's (1985) monumental study of Seri ethnobotany for its many examples.

The Midriff Islands figure prominently in Seri oral history. Their traditions provide a unique view of the human history of Isla Tiburón from the Seri perspective (for example, Herrera 1988), and they are the definitive source of information about the people of Isla San Esteban (Bowen 2000:5-30; Moser 1963). They have proved to be generally reliable sources for specific events over a span of at least 120 years; thus material recorded in the mid-twentieth century may provide credible accounts of events as far back as the early nineteenth century (Bowen 2000:397-406; Felger and Moser 1985:171). It is possible that Seri oral traditions preserve a general record of much more ancient events, although interpreting them requires an understanding of Seri narrative conventions (Bowen 1976:108, 2000:404-406; Hills and Yetman 2007:512-518).

The Archaeological Record

Not surprisingly, most islands where Indians were documented by Europeans have an archaeological record of their activities.

Fortunately, Indians also left archaeological evidence of their activities on many islands for which there are no known written records. This section describes the major types of sites, artifacts, and features (see accounts of individual islands in Chapter 3 for illustrations).

Percussion-flaked stone implements and manufacturing debris are by far the predominant artifacts on all the islands, and the vast majority of lithic remains are waste flakes and cores. Formal tools, made to a predetermined shape, are scarce. Apparently, overall form was seldom important, and the objective of most tool-making was to produce serviceable working edges for a variety of tasks including cutting, scraping, chopping, planing, and gouging. Thus tools are often nothing more than retouched flakes and cores, or rocks with two or three flakes removed to produce a crude working edge. Many tasks may have been performed with expedient tools—simple flakes, cores, or even just broken rocks with sharp edges, produced on the spot, used briefly without further modification, and then discarded. As a result, in the field it is often impossible to distinguish barely-used flakes and cores from debitage.

In the Midriff region, two distinctive kinds of tools were produced by minimal edge modification. Agave knives (mescal knives) were usually made from tabular slabs by detaching a few large flakes from one or two edges to produce a ragged cutting edge. Denticulates are characterized by a pointed beak or projecting "tooth," usually chisel-shaped or rounded by retouching, and many specimens have two or even three such projections. Presumably they were used as gouging tools.

Formal artifacts consist mainly of scrapers and projectile points, some of which were finished by careful retouching. Scrapers are oval to round in outline and characteristically thick, with one more or less flat face and the other face steeply convex, creating a "domed" appearance. They may be flaked unifacially or bifacially, and some have been retouched around part or all of the edge. Some domed implements may be essentially retouched cores, but most appear to have been designed as tools. The term "scraper," of course, is merely a label—they may well have been multipurpose implements.

Projectile points are widespread but generally few in number. They have been found on Islas Cerralvo, Espíritu Santo/Partida Sur, probably Los Coronados and San Marcos, and on Tiburón, San Esteban, and Ángel de la Guarda. They range from large lanceolate points and stemmed forms to small triangular points with concave bases. They also include small leaf-shaped specimens, some of which might be unfinished points, and large specimens that were probably used as knives.

Large bifaces are common on a few of the larger islands. Most have a rounded base and a slightly pointed tip, and they display various degrees of reduction and thinning. They generally occur at quarry-workshop sites, and the majority of these are broken. In many cases bifaces are probably performs—stages in the manufacture of projectile points and knives—and were abandoned because of breakage or inability to proceed further. Others, however, both broken and whole, occur far from manufacturing sites, suggesting that some of these seemingly unfinished artifacts may have been put to use as tools.

Besides flaking, rocks were modified by grinding and pounding. Ground stone tools throughout the islands are predominantly unshaped water-worn rocks that were used as metates and manos. Many show heavy wear, suggesting the importance of plant foods in the aboriginal diet. Some metates have dark organic residue or red pigment adhering to the grinding surface. Manos often have battered ends, indicating ancillary use as hammers, and some islands have hammers that never saw service as manos.

Many islands have large shells of the clam *Dosinia ponderosa* that were flaked for use as cutting or scraping tools, and these often have edges dulled by heavy use. Some sites include large unmodified scallop shells (*Lyropecten subnodus*) that may have served as scoops or dippers. Native pottery is known only on islands in the Seri region, although some post-European pottery has been found on Isla Espíritu Santo.

Discarded unmodified shells and animal bones, indicative of diet, are abundant on some southern islands, but are rare or absent on all but the largest Midriff Islands. Similarly, hearths occur on some larger islands, but many smaller islands have no evidence of fire and therefore no charcoal that could be used for radiocarbon dating.

On some islands, structures predominate over artifacts. These include circular to oval clearings and *corralitos*. Corralitos (rock rings) are enclosures of piled rocks, usually with an opening on one side, and sometimes with an interior cleared of large rocks. Most islands also have stone circles, which consist of individual rocks set on the ground in a circular or oval pattern. Generally the rocks in these circles are widely spaced but on some islands they are close-set and may touch or nearly touch one another. Among the most widespread structures are rock cairns (mounds of large rocks piled two or more high) and some islands have rock clusters (more or less circular aggregations of unpiled rocks). Other structures include rock alignments, ground figures, and talus pits (excavated pits in talus slopes). It is probably safe to assume that most clearings and corralitos served as sleeping surfaces and windbreaks, but we have essentially no idea how the other types of structures were used.

Many islands have few or no occupation areas that can reasonably be considered "sites." In these cases, archaeological remains occur almost entirely as isolated artifacts and structures, suggesting small groups or lone individuals ranging widely over the landscape during brief visits. On other islands, recognizable habitation sites include open-air camps, rockshelters, and shell middens. Specialized sites are mostly quarry-workshop areas where tool rock was obtained and flaked tools were roughed out or manufactured. These range in scale from single knapping events to sites with tens of thousands of waste flakes and spent cores. Other specialized sites, especially those containing structures but no artifacts, are of unknown but possibly esoteric function. They include long lines of rock cairns, sometimes with hundreds of individual structures, concentrations of rock clusters and alignments, groups of stone circles, talus slopes with dozens to hundreds of pits, clusters of corralitos on high summits, and sites with a mix of structures. Rock art is known only from Islas Espíritu Santo/Partida Sur, Tiburón, and San Esteban, and only Islas Cerralvo, Espíritu Santo/Partida Sur, San José, and Tiburón have produced burials.

Despite the idiosyncracies in the archaeological assemblages of each island, three broad generalizations can be made. First, most island assemblages are simplified versions of those on the adjacent coast, reflecting the fact that (in historical times, at least) they were the product of the same people, who moved back and forth freely. Secondly, Indians on both sides of the Gulf were known historically as pragmatic people who seldom invested more effort in a task than necessary, and this is born out by the basic simplicity of island archaeological assemblages throughout the Gulf. Lastly, most types of structures and artifacts are widespread and strikingly similar wherever they occur. Thus corralitos, rock cairns, and rock clusters on Isla Espíritu Santo in the southern Gulf look pretty much the same as those on Isla Tiburón and Ángel de la Guarda in the Midriff region, and stone circles on Isla Santa Catalina are essentially similar to those on Islas Salsipuedes and San Luis. Core choppers, metates, and manos are similar everywhere, and historic accounts leave no doubt that the cane balsa, the all-important watercraft that enabled Indians to reach the islands, was similar throughout the Gulf. People may have differed in language and ideational aspects of culture, but they exploited island resources with much the same technology.

Archaeological remains on the islands constitute tangible proof of native people. But it is important to note that an absence of remains does not necessarily imply an absence of people. Many important activities documented historically or ethnographically required no equipment or none that would be recognizable archaeologically. These include harvesting most plant foods, gathering seabird eggs, fishing with sharpened wooden spears, hunting sea lions with thrown rocks, hunting pelicans with bare hands and stealth, pulling chuckwallas from their burrows, and collecting a wide variety of slow game such as rodents, small birds, small lizards, snakes, insects, larvae, crustaceans, and shellfish. Faunal remains other than shells may not be preserved at open surface sites, nor artifacts made of perishable materials; yet many islands have no stratified deposits or rockshelters where fragile remains such as these might be found. Perhaps the quintessential perishable artifact is the cane balsa. Although historical records document their use in large numbers throughout the Gulf over a span of nearly 400 years, only two specimens are known to have survived intact beyond the early 1920s, and only because they were collected as ethnographic specimens.

Even non-perishable remains can disappear without a trace. Massive erosional events, which are frequent consequences of tropical storms, can wash away or bury entire sites on islands with steep topography, and wave action generated by storm surges can carry shoreline sites out to sea (Fig. 2.1). Plant growth and animal activity can move artifacts and destroy structures by displacing their component rocks. Modern human visitors, whether panga fishermen, scientists, or tourists, sometimes pick up artifacts, and site looters have been hard at work on some islands. These nefarious individuals are capable of stripping clean small islands with few artifacts, thereby obliterating the entire record of native people (Bowen 2004:201-203).

Systematic archaeological surveys on Gulf islands are recent. In the southern Gulf, extensive surveys and excavations have been conducted on Islas Cerralvo and Espíritu Santo/Partida Sur since the mid-1990s (Fujita 1998, 2006, 2008a; Fujita and Poyatos de Paz 1998). In the Midriff region, Islas Tiburón and San Esteban were surveyed in the 1970s and 1980s (Bowen 1976, 2000; Villalpando 1989). Since 2004, the other large Midriff Islands and many small ones have been surveyed (Bowen:field notes). Many Gulf islands, however, have never been systematically investigated. On some of these, scientists from other fields and even casual visitors have recorded cultural material, and their observations are included in the accounts of individual islands that follow. Others, however, including the large islands of Santa Cruz, Monserrat, and Tortuga, remain completely unknown archaeologically.

WATER RESOURCES

For native people, the habitability of the Gulf islands depended in part on the availability of fresh water. Permanent habitation required permanent springs or *tinajas* (bedrock pools). Seasonal occupation required long-term water sources coupled with a reliable rainy season. Both conditions were met in the southern Gulf, as was clearly articulated as early as the 1640s by Pedro Porter y Casanate:

Some of [the islands in the southern

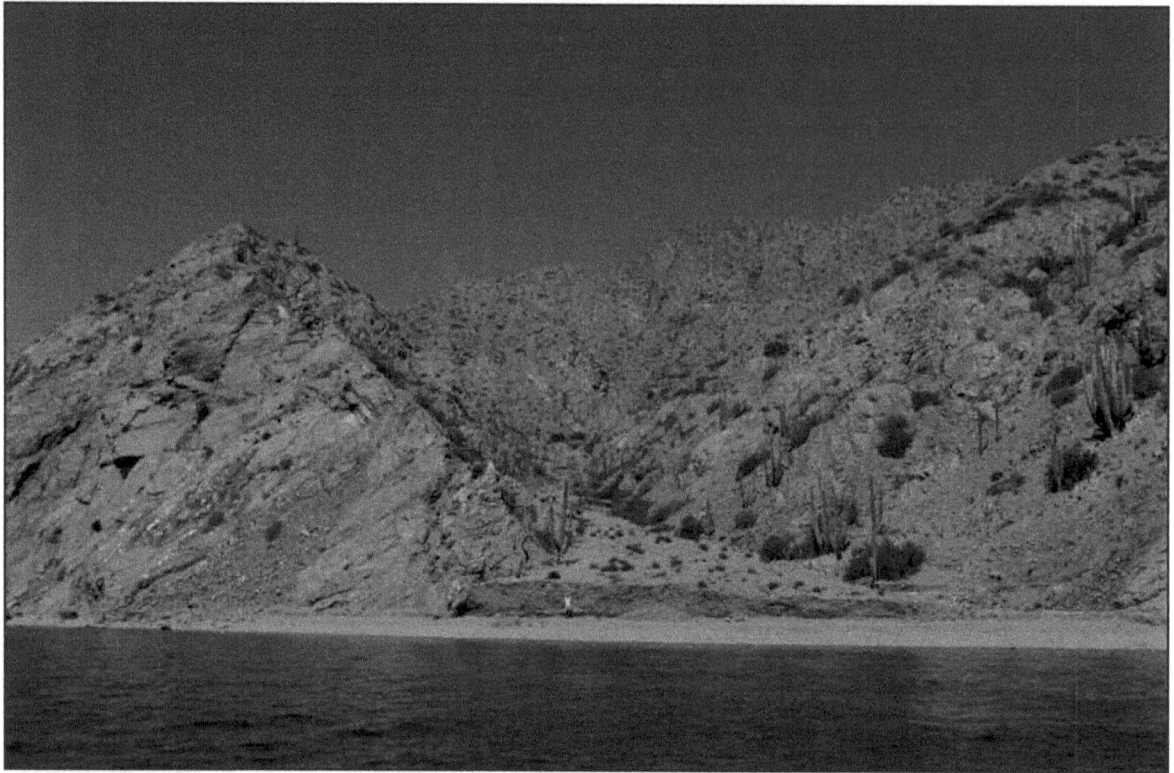

Figure 2.1. Isla San Lorenzo, western side. An example of how natural processes can eradicate evidence of human occupation. Four months before the photo, heavy rains from Tropical Depression Javier scoured the hillsides in this valley and redeposited loose material in the arroyo to a depth of some 2.5 m. Any cultural material on the slopes would probably have been washed into the arroyo, and any sites in the arroyo not carried out to sea are now deeply buried. The person in the lower center of the photo, 6 ft 2 in tall (185 cm), stands on the pre-storm outwash surface. Photograph looks northeast. January 2005.

Gulf] are permanently inhabited while others are inhabited by peninsular people only during the rainy season, which occurs during the months of August, September, and October (Porter 1970:892).

The Midriff region receives most of its sparse precipitation in the late summer and early winter. But rainfall in this region is characterized far more by its extreme variability and unreliability than its seasonality, and a year or more may pass in any given locality with no precipitation at all (Cavazos 2008: 67-73; Cody and others 2002:67-68). Although Isla Tiburón (and possibly Isla San Esteban) had permanent water (Fig. 2.2) and was apparently inhabited by Seris throughout much of the year, exploitation of other Midriff islands must have been opportunistic rather than seasonal.

Whenever the sporadic rains filled the tinajas, islands without permanent water would have been temporarily habitable (Fig. 2.3).

What is less obvious is that native people throughout the Gulf were quite capable of exploiting islands with no water resources whatsoever. By bringing water with them or subsisting temporarily on water substitutes, Indians could make short-term visits to completely waterless islands. The Seris transported water in pottery vessels both on land and at sea (Felger and Moser 1985:80-81, 131) while both Seris and peninsular people used the stomachs and bladders of various animals as canteens (Aschmann 1959:59; Felger and Moser 1985:81). Both groups were adept at supplying their fluid requirements from columnar cactus fruits and other plants, and they

Figure 2.2. Isla Tiburón, southwestern side. One of the permanent spring-fed tinajas at the Sauzal waterhole. January 1983.

sometimes relied on these water substitutes for extended periods (Bowen 2000:410-416). Nearly all Gulf islands have at least one species of columnar cactus that yields fluid-rich fruit that could be substituted for water. The Seris knew how to make this juice even more palatable by fermenting it into wine (Felger and Moser 1985:247).

Water sources may change over time. Springs can dry up or reappear in other locations. Changes in vegetative cover can extend or reduce the longevity of tinajas. Most importantly, changing climate conditions can alter an island's water supply. The water resources we see during today's warming regime largely determine our perception of an island's habitability, but these resources may not be the same as existed 100 years ago, 1000 years ago, or 10,000 years ago.

As for the documentary record of water

resources, it is important to recognize that historical statements about water reflect the cultural perceptions of those making them and may not be as literally true as the words suggest. Some observers have considered islands waterless not because there was no water at all, but because water was not abundant or permanent. Thus early Spanish explorers sometimes declared islands waterless if they saw no flowing water or if the flow was insufficient to support agriculturally-based settlements of Europeans. Even more recent writers have expressed this bias:

> In the northeastern corner [of Isla Ángel de la Guarda] lies a quite pretentious valley through which a temporary stream runs immediately after rains. Rumors and tales of flowing water and weird occupants circulate all over the Gulf....

Figure 2.3. Isla Ángel de la Guarda, western side. Three well-shaded tinajas photographed three months after the last known rain. Their combined water volume at this time was about 3050 liters; maximum capacity is about 6450 liters. December 2006.

The hard fact remains, however, that La Guardia [sic] is a desert, uninhabited and unwatered (Bancroft 1932:354).

But quantity does matter. A pothole that holds a few dozen liters for a few days after a rain cannot support a population, but it can supply the needs of a small party of native people on a brief visit. In fact, historic assertions that an island is waterless are seldom trustworthy, and even twentieth century statements should be regarded with caution. Some statements have been mere inferences based on the arid appearance of the landscape as seen from a passing ship. Thus Father Juan María de Salvatierra, sailing past Isla San Esteban in 1709, could not believe the island had water even though Seri Indians had assured him that it did (Bowen 2000:62). Other statements have been generalizations made from insufficient exploration. In 1765 Father Wenceslaus Linck inferred incorrectly that Isla Ángel de la Guarda was waterless after a scant three days on that huge island (Barco 1967:27). Some assessments are strangely perplexing. In 1746, William Strafford (Guillermo Stratford) declared Isla Espíritu Santo waterless even though, as a pilot for the Jesuits with long experience in the southern Gulf, he must have known that both Indians and Spanish pearl hunters obtained water there (Stratford 1958:54). And sometimes knowledge of water sources is lost. The eighteenth century Jesuits knew that Isla Cerralvo had fresh water (for example, Taraval 1931:243) but Edward Nelson and Edward Goldman, after two days there in February 1906, concluded that the island was "waterless and uninhabited" (Nelson 1922:48). Similarly, in 1930 Griffing Bancroft, who motored past Cerralvo without bothering to go ashore, declared it "uninhabited, unwatered, and unused" (Bancroft 1932:180). These misperceptions of Cerralvo were finally laid to rest in 1962 by Richard Banks:

> Previous [scientific] discussions of the island report that it is without fresh water...but there are several small springs at the heads of arroyos which provide limited quantities of good water (Banks 1962:117-118).

Discoveries of water are still being made within the scientific community. Although the small island of San Pedro Nolasco had been reported as waterless (Felger and Lowe 1976:21; Gentry 1949:96), in 2007 botanists Richard Felger and Benjamin Wilder learned that fishermen and local researchers have known of a permanent seep there for some time, located less than 100 m from the landing where nearly every party of scientists has gone ashore (Felger and Wilder 2009). The lesson, of course, is that one can only be sure an island is waterless after conferring with everyone who goes there and inspecting every square meter.

NATIVE WATERCRAFT

Native people could not have exploited the Gulf islands without seaworthy watercraft. Canoes were known in the southern peninsula, but the vessel used everywhere in the Gulf was the cane balsa (Heizer and Massey 1953). These craft were made of three bundles of *carrizo* or reedgrass (*Phragmites australis*) tied together and, like kayaks, were propelled with double-bladed paddles. Most were designed for one or two people, but some could accommodate three or four (Correa 1946:554; McGee 1898:218; Vizcaíno 1930:211). The Seris of Isla San Esteban made even larger balsas capable of holding three families—some six adults and nine or ten children (Bowen 2000:22; Felger and Moser 1985:310). It is said that the *Hant Ihiini* Seris, who lived on Baja California, made giant balsas capable of transporting as many as 20 or 25 men (Stephen Marlett, personal communication 2008).

Balsas disappeared from the southern Gulf with the demise of the peninsular Indians late in the eighteenth century, but the Seris used them throughout the nineteenth century. A few Seri balsas remained in use up to the early 1920s (Davis 1922:32; Sheldon 1921: 137) and an abandoned and rotting balsa was seen as late as the 1930s (Felger and Moser 1985: 311).

The Seris still remembered much about balsas and balsa travel in the latter part of the twentieth century, and this knowledge provided the basis for the best general account of these watercraft (Felger and Moser 1985:310-311). Remarkably, in 1978, linguist Mary Beck Moser was able to record the childhood memory of an elderly Seri woman, María Antonia Colosio, who travelled by balsa from Isla Tiburón to Isla San Esteban in the first decade of the twentieth century. This voyage of nearly 12 km traversed one of the most treacherous channels in the Gulf. As María Antonia recalled:

> We were on two balsas that were tied together, side by side....They had put blankets and water on, jugs as big as this [gesturing, indicating a large water vessel]. Since the jugs were full of water, they were tied in place. Plants were stuck in the mouth of the jugs, and the jugs were in carrying nets. I didn't want to get on, but my father caught me and put me on. After he caught me and put me on, he tied me behind a blind man who went along to paddle. Then I cried a lot, but he didn't pay any attention to me. That's how we went to *Coftécöl* [Isla San Esteban]. It was so dangerous when we almost entered the area called *Ixötáacoj* [Big Whirlpool]. The sea just swirled and churned. The wind wasn't blowing but the water was choppy. It just churned, it was danger-

ous. The sea was going around. Everything just roared. The children and old women all cried. The old man Pozoli just said, "We'll land really soon." As we were going to land, he sang to the shore. And it seemed we landed right away. The men paddled with all their strength, and we landed near the rocks (Felger and Moser 1985:131-132).

Outsiders have always been intrigued by the balsa, and many of them recorded what they saw. A balsa was sketched in 1873 (Fig. 2.4) and at least two balsas were photographed, one about 1896 (Fig. 2.5) and the other in 1922 (Fig. 2.6; Davis 1965:142). Most observations, however, have been recorded in writing, each with a slightly different perspective.

In 1539, Francisco de Ulloa described a Cochimí balsa on the peninsular side of the Midriff region:

> They had a little raft which they must have used in fishing. It was made of canes tied in three bundles, each part tied up separately, and then all tied together, the middle section being larger than the laterals. They rowed it with a slender oar, little more than half a fathom long [about 1 m], and two small paddles, badly made, one at each end (Ulloa 1924:327-328).

In 1615 Nicolás de Cardona wrote of balsas in the southern Gulf near La Paz:

> Their boats are of three bundles of thin cane, two on the sides and one in the center, very well tied together so that in each of these vessels two persons may ride....In each boat an Indian goes out to fish, rowing with both hands with a two-bladed oar (Cardona 1974:100).

Writing in 1633 of the Indians around La Paz, Francisco de Ortega observed that:

Figure 2.4. Seri paddling a balsa near Isla Tiburón, December 1873, as sketched by W.F. Beardslee, ship's artist aboard the U.S.S. Narragansett. *Reproduced from McGee (1898:Figure 28).*

Figure 2.5. Seri balsa collected on the Sonoran coast in December 1895 by WJ McGee, photographed on the grounds of the Smithsonian Institution by William Dinwiddie. Reproduced from McGee (1898:Plate 31).

Figure 2.6. Seri balsa, May 1922. Ramón Blanco on his balsa with a harpoon in a turtle-hunting pose, Bahía Kino, Sonora. His double-bladed paddle lies in front of his left foot, and the craft would have been paddled from a kneeling position. This balsa, probably the last one used in the entire Gulf, was purchased by Edward H. Davis as an ethnographic specimen. Photo by Edward H. Davis, May 1922. Courtesy National Museum of the American Indian (No. 24086).

they always go out to sea in the summertime when they take their sustenance from the sea, travelling in little balsas which they take four to six leagues [10 to 15 km] out from shore (Ortega 1970b:449).

In 1648 and 1649, Pedro Porter y Casanate saw balsas off the southern peninsula:

the Indians use boats made of five poles joined together [and] others made of carrizo cane, which they propel with great speed and use to cross over from the peninsula to the Gulf islands (Porter 1970:892).

In 1746 William Strafford remarked that in front of Bahía de Cerralvo, just south of La Paz:

there is an island at a distance of three leagues [8 km] to the east, also named Serralbo....The Indians from the aforementioned bay inhabit it seasonally, reaching it in balsas, which they make from carrizo or cane, in order to obtain water, agave, and sea turtles with which to sustain themselves (Stratford 1958:54).

In the Midriff region, the earliest description of Seri balsas was penned by Father Adam Gilg in 1692:

My Seris do not build their boats of planks, but of three bundles of reeds bound together, which are joined in a narrow point behind and before, and are widely separated from each other in the middle, forming a hollow bilge. In place of an anchor, they throw a good sized stone into the depths when they want to stop (Gilg 1965:50-51).

In 1750, Manuel Correa wrote that the balsas he saw on Isla Tiburón were:

composed of many small carrizos arranged in three bundles, thick in the middle and thin on the ends, joined together with plaited cord. These balsas were generally five or six varas long [4.2 m to 5.0 m] and about one and one-quarter varas wide [1.0 m] in the middle, tapering proportionately toward the tips. These balsas support the weight of three or four people and they slice through the water very rapidly. The paddles are wooden shafts about two varas long [1.7 m] with two blades fastened to the tips, and the paddler, holding it in the middle, paddles from one side to the other (Correa 1946:554).

Off Isla Tiburón in 1826, Robert Hardy:

observed about fifteen or twenty canoes [that is, balsas], made of three long bamboo bundles fastened together, which terminated in points at the head and stern. From their natural buoyancy they easily support the weight of an Indian, although the water penetrates through the sticks in every direction. When loaded, the centre sank down a little below the water's edge, the bow and stern only rising about six or eight inches above it (Hardy 1977:291-292).

In 1875 the crew of the U.S.S. *Narragansett* saw Seris on balsas in the vicinity of Isla Tiburón:

They are made of long reeds, which are bound together with strings after the

manner of fascines, three of which... have sufficient buoyancy to support one or two persons. They kneel in these canoes when paddling, the water being at the same level in the canoe as outside of it (Belden 1880:145).

The most detailed physical description was written by WJ McGee, who collected a Seri balsa in 1895. This specimen was:

made of three bundles of carrizal or cane lashed together alongside, measuring barely 4 feet abeam, 1½ feet in depth, and some 30 feet in length over all....The finished balsa is notably light and buoyant. The Boca Infierno specimen was estimated to weigh about 250 pounds (113 kilograms) when thoroughly dry, and little more than 300 pounds (126 kilograms) when completely wet; so that it could easily be picked up by three or four, or even by two, strong men and carried ashore to be hidden in the fog-shrubbery skirting the coast (McGee 1898:216-218).

McGee's party actually had first-hand experience with this balsa at sea, paddling it along the Sonoran coast to a location where it could be hauled inland for its long journey to the Smithsonian. McGee was dazzled by its performance:

The craft floated high with one man aboard, rode better with two, carried three without much difficulty even in a fairly heavy sea, and would safely bear four adults aggregating 600 pounds (272 kilograms) in moderate water. The most striking features of the craft afloat are its graceful movement and its perfect adaptation to variable seas and loads....Almost equally striking features of the balsa are its efficiency and safety under the severe local conditions.

Carrying twice its weight of (chiefly) living freight, it breasts gales and rides breakers and stems tiderips that would crush a canoe, swamp a skiff, or capsize a yawl; while if caught in currents or surf and cast ashore it is seldom wrecked, but drops lightly on beach or rocks, to be pushed uninjured by the broken wave-tips beyond the reach of pounding rollers, even if it is not at once caught up by its passengers and carried to complete safety. The strength of the craft is amazing (McGee 1898:218).

Another Seri balsa, also obtained as a specimen, was one of the last ones in actual use. It was purchased at Bahía Kino in 1922 by Edward H. Davis, who recorded the ordeal of beaching it and loading it on his truck:

The balsa was very heavy and some distance away [from the truck]. It also had some things tied on it. So Bob [Roberto Thomson Encinas] pulled off his pants. We pushed it into the surf and [Bob] first stood up but it rocked, so he got on his knees and pushed out with a pole and got it headed to the rollers where it rode easier. He then lost his pole and in struggling to recover it, the balsa rolled over with him and dumped him in the surf. But, he grabbed the side and it rolled upright again. He could then touch bottom, so he came in through the surf with it and we dragged it up and unloaded the three turtle shells and a few other things. There were 7 of us and we put oars under it and gradually lifted it and carried it to the truck. It must have weighed 600 or 700 lbs a good part of which was water. I must say I was glad to see it started on its long land journey (Davis 1922:32).

In short, balsas were the standard watercraft throughout the Gulf from the time Europeans first appeared on the scene, and they served the native people well for both subsistence and transportation.

Chapter 3
The Record of Individual Islands

This chapter documents evidence, or lack thereof, for indigenous people on 32 Gulf of California islands (Table 3.1). The list includes all 22 islands larger than 2 km² plus 10 smaller islands for which some data exist (Table 3.2). The order of presentation is southeast to northwest, approximating the chronological order in which the islands became known to Europeans.

The information for individual islands that follows should not be construed as comprehensive island histories or archaeological reports, but simply as representative examples or summaries of currently available evidence of native people. For documents that give distances in Spanish leagues, the metric equivalent of the league can be considered to be 4.2 km (Barnes and others 1981:71; Chambers 1975).

ISLA CERRALVO

Water Resources

Father Sigismundo Taraval, writing about 1737, stated that:

> The island of Cerralvo lies about six or seven leagues off the coast [of Baja California] and sixteen from the port [of La Paz]. Approximately triangular in shape, it is about twelve leagues long. Of its two water-holes, only one has good water (Taraval 1931:243).

Based on his 1960 field work, ornithologist Richard Banks remarked that:

> Previous discussions of the island report that it is without fresh water...but there are several small springs at the heads of arroyos which provide limited quantities of good water (Banks 1962:117-118).

Botanist George Lindsay wrote of his 1962 visit that:

> We found two small springs in the arroyo [on the northwest side of the island], shaded by giant fig trees (*Ficus palmeri*). The spring water was potable but had a sweet taste (Lindsay 1962:31).

According to herpetologist Lee Grismer:

> On the southwest end of Isla Cerralvo, the steep rocky foothills meet the low coastal dunes along the back edges of the beaches, forming a long, narrow basin in which rainwater and runoff collect and provide breeding pools [for the island's two toads]. *Bufo punctatus* [the Red-Spotted Toad] also breeds in a small, palm-lined canyon with semi-permanent water at the northeastern end of the island (Grismer 2002:71).

Table 3.1. Locations of the islands, compiled from Google Earth satellite imagery (http:\\earth.google.com). Latitude and longitude figures, rounded to the nearest 5", indicate the approximate centers of the islands. Distances from islands to nearest mainland and nearest major island are shortest straight line measures unless otherwise noted

Island	Latitude (N)	Longitude (W)	Nearest Mainland	Distance (km)	Nearest Major Island	Distance (km)
Cerralvo	24°14'30"	109°53'00"	Baja California Sur	12.0		
Espíritu Santo	24°28'10"	110°20'10"	Baja California Sur	6.3	Partida Sur	0.02
Partida Sur	24°33'40"	110°23'10"	Baja California Sur (via the western coast of Espíritu Santo)	23.0	Espíritu Santo	0.02
San Francisco	24°49'50"	110°34'30"	Baja California Sur	7.4	San José	2.5
San José	24°59'00"	110°37'10"	Baja California Sur	4.8		
Santa Cruz	25°17'05"	110°43'05"	Baja California Sur	19.8		
Santa Catalina	25°39'10"	110°46'55"	Baja California Sur	25.1	Monserrat	20.8
Monserrat	25°40'55"	111°02'00"	Baja California Sur	13.7	Carmen	18.1
Danzante	25°47'15"	111°15'05"	Baja California Sur	2.6	Carmen	2.8
Carmen	25°57'10"	111°10'05"	Baja California Sur	6.5		
Los Coronados	26°07'15"	111°16'25"	Baja California Sur	2.7	Carmen	11.2
San Ildefonso	26°37'50"	111°25'45"	Baja California Sur	10.1		
San Marcos	27°13'05"	112°04'15"	Baja California Sur	4.9		
Tortuga	27°26'30"	111°52'45"	Baja California Sur	35.6	San Marcos	26.9
San Pedro Nolasco	27°58'00"	111°22'40"	Sonora	14.8		
San Pedro Mártir	28°22'50"	112°18'30"	Baja California	50.1	Dátil	35.4
			Sonora	51.5	San Esteban	39.5
Alcatraz	28°48'40"	111°58'05"	Sonora	1.5		
Tiburón	28°58'40"	112°21'25"	Sonora	1.6		
Dátil	28°43'20"	112°17'25"	Sonora	29.4	Tiburón	1.7
Patos	29°16'15"	112°27'30"	Sonora	7.6	Tiburón	8.9

Table 3.1. Locations of the islands, cont'd

Island	Latitude (N)	Longitude (W)	Nearest Mainland	Distance (km)	Nearest Major Island	Distance (km)
San Esteban	28°41'50"	112°34'40"	Baja California	34.6	Tiburón	11.7
			Sonora (via southern tip of Tiburón)	53.1	San Lorenzo	16.8
San Lorenzo	28°37'50"	112°49'05"	Baja California	16.3	Las Ánimas (at low tide)	0.1
					San Esteban	16.8
Las Ánimas	28°41'40"	112°54'50"	Baja California	20.4	San Lorenzo (at low tide)	0.1
Salsipuedes	28°43'30"	112°57'20"	Baja California	19.3	Las Ánimas	1.5
Rasa	28°49'25"	112°58'50"	Baja California	20.8	Partida Norte	8.4
					Salsipuedes	9.6
Partida Norte	28°53'25"	113°02'20"	Baja California	17.9	Ángel de la Guarda	12.3
Cardonosa Este	28°53'10"	113°01'45"	Baja California	19.3	Partida Norte	0.7
Ángel de la Guarda	29°13'50"	113°19'25"	Baja California	12.8		
Estanque	29°03'55"	113°05'20"	Baja California (via the southern tip of Ángel de la Guarda)	28.9	Ángel de la Guarda	0.6
Mejía	29°33'20"	113°34'15"	Baja California	24.2	Ángel de la Guarda	0.7
Piojo	29°01'00"	113°27'55"	Baja California	6.4	Coronado	2.4
Coronado	29°03'50"	113°30'30"	Baja California	2.3		
San Luis	29°58'15"	114°24'30"	Baja California	5.2		

Documentary Record

Reporting on his 1632 visit, shipwright and pearl hunter Francisco de Ortega wrote:

> we arrived at Isla de Cerralbo where we dropped anchor and went ashore....On the shore facing the peninsula there is a great bay, and at this bay there is a large community of Indians (Ortega 1970a:420).

In 1721, Father Ignacio María Nápoli concluded that a group of Indians he encountered at La Paz:

> are really the Indians who live in the island of Serralvo, and who are accustomed to visit the bay of Las Palmas (Nápoli 1970:70).

In 1746, William Strafford, longtime pilot for the Jesuits, wrote that a bay on the peninsula was named "Serralbo," and that:

in front of this bay there is an island at a distance of three leagues to the east, also named Serralbo, which is three or four leagues in length, running from southeast to northwest. The Indians from the aforementioned bay inhabit it seasonally, reaching it in balsas, which they make from carrizo or cane, in order to obtain water, agave, and sea turtles with which to sustain themselves (Stratford 1958:54).

Although Isla Cerralvo had been largely abandoned by the 1720s, rebel Pericú Indians retreated there during the rebellion of 1734-1737. Father Taraval wrote that in 1736:

the commander [Don Manuel Huidobro] decided to send twenty-five soldiers and fifty archers in the launch and a canoe over to the island of Cerralvo where, according to reports, several ringleaders of the Pericúes were hiding....This island has been to the Pericúes of Mission

Table 3.2. Islands in order of size. Based on Murphy and others (2002: Table 1.1-1) and Carabias and others (2000:199)

Islands Larger Than Two Square Kilometers		Islands Smaller Than Two Square Kilometers	
Island	Area (km^2)	Island	Area (km^2)
Tiburón	1223.5	Partida Norte	1.4
Ángel de la Guarda	936.0	San Ildefonso	1.3
San José	187.2	Dátil	1.3
Carmen	143.0	Salsipuedes	1.2
Cerralvo	140.5	Estanque	1.0
Espíritu Santo/Partida Sur	106.9	Rasa	0.7
Santa Catalina	41.0	Piojo	0.6
San Esteban	40.7	Alcatraz	0.5
San Lorenzo	33.0	Patos	0.5
San Marcos	30.1	Cardonosa Este	0.1
Monserrat	19.9		
Santa Cruz	13.1		
Tortuga	11.4		
Coronado (Smith)	9.1		
Los Coronados	7.6		
San Luis	6.9		
Danzante	4.6		
San Francisco	4.5		
Las Ánimas	4.3		
San Pedro Nolasco	3.5		
San Pedro Mártir	2.9		
Mejía	2.3		

Santiago...a place of refuge for the most dangerous culprits (Taraval 1931:243).

Archaeological Record

Archaeological investigations on Isla Cerralvo began in 1903 when mining engineer and naturalist Léon Diguet recovered burials assignable to the Las Palmas funerary complex (Diguet 1973; Massey 1955:49; Reygadas 2003:34). In 1997 archaeologist Harumi Fujita spent a month conducting a comprehensive survey of Cerralvo's rugged coastline where she recorded four large shell middens (Fig. 3.1) and a fifth site near the island's summit (Fujita 1998).

The midden sites contain a variety of simple flaked stone tools including choppers, scrapers, knives, and projectile points, along with associated debitage and hammerstones.

Other artifacts include metates, manos, and flaked bivalve shells. The largest site (Site 1) covers approximately 4.8 hectares and includes three circular clearings and two hearths. All four middens produced shells (some 25 species), fish bones, and bones of sea turtles and dolphins. A burned deer bone at one site suggests contact with the peninsula. Four radiocarbon samples from a test pit at Site 2 produced calibrated and corrected dates ranging from AD 701 (1 sigma range AD 560-900) to recent (from a surface sample) (Fujita 2008b). Site 5, near the summit of the island, contains a hearth, a large bifacially-flaked knife (Fig. 3.2), a metate, three hammerstones, and most of the same flaked stone assemblage, but relatively few shells, most of them fragmentary. Fujita (1998:26-33, 68) interprets all five sites as open-air camps.

Figure 3.1. Isla Cerralvo, southwestern side. A small portion of a shell midden 300 m long and 40 m wide. Artifacts include flakes and cores, two metates, rocks showing traces of use, and burned rocks. Shells include nine species of bivalves and 13 species of gastropods. Photograph looks northeast. October 1997. Photograph courtesy of Harumi Fujita and Instituto Nacional de Antropología e Historia.

Figure 3.2. Isla Cerralvo. Bifacially-flaked knife of local rhyolite from an open-air camp at more than 700 m elevation near the island's summit. Length is 18 cm, width 11 cm, thickness 4 cm. October 1997. Photograph courtesy of Harumi Fujita and Instituto Nacional de Antropología e Historia.

ISLA ESPÍRITU SANTO AND ISLA PARTIDA SUR

Apparently, these two islands separated from each other only recently. For most of the historic period, they were a single land mass referred to as "Espíritu Santo." In the 1870s, they were still a single island, but by then the name "Partida" had been introduced:

> At this point [on the eastern side] the general coast line is broken by an indentation 1 3/4 miles deep and a mile wide at its outer part, which, with a corresponding indentation on the western side, nearly divides the island....Isla Partida is the name given to that portion of the island lying north of the two indentations or coves just spoken of. It is joined to the main part of the island by a neck of land less than 300 yards wide and of moderate height (Belden 1880:77).

Eighty years later, herpetologist Frank Cliff found that the two land masses were separated at low tide "by a water gap of only ten feet" and that, despite the separation, "These islands are a biological unit" (Cliff 1954:70). Today, the two islands are separated by a channel 20 m wide (Harumi Fujita, personal communication 2008). However, since they were a single physiographic and cultural entity in the historic past, in this paper it is often appropriate to treat them as a single unit under the combined name "Isla Espíritu Santo/Partida Sur."

Water Resources

Francisco de Ortega, who visited Isla Espíritu Santo in 1632, reported that:

> at the southern end there is another port where we anchored for two days and where we found fresh water. After taking on what we needed, we left this port (Ortega 1970a:423).

In 1734, when the Pericú Rebellion broke out in the Cape region, Father Sigismundo

Taraval and some loyal Indians fled from the peninsula to:

> the island of Espíritu Santo. The Pericúe Indians knew where to find water, since they...knew intimately nearly all the neighboring islands (Taraval 1931:70).

In 1883, Dutch anthropologist Herman ten Kate wrote that Ensenada de la Candelera:

> is one of the few spots on Espíritu Santo where fresh water is found (Kate 1977:58).

Naturalists Edward Nelson and Edward Goldman visited Isla Espíritu Santo in February 1906 in conjunction with their survey of Baja California's natural resources. They reported that:

> Espíritu Santo has no fresh water except in natural tanks after rains (Nelson 1922:91).

Lee Grismer listed two toads on both Espíritu Santo and Partida Sur, both of which require seasonal supplies of fresh water to reproduce (Grismer 2002:70, 83).

Documentary Record

In 1596, Gonzalo de Francia, who served as a boatswain with Sebastián Vizcaíno, stated that Espíritu Santo was inhabited seasonally:

> At the entrance of the port [of La Paz] there is an island which we called "Isla de Mujeres." It is uninhabited as people only go there in summer in some small cane *balsas* (Francia 1930:219).

When Francisco de Ortega anchored at Isla Espíritu Santo in 1632:

> many Indians who are settled at a port on the northwestern coast of this island came out to the ship, and at this port

there are many great shell middens from the [shellfish] the Indians eat....We left this port to reconnoiter the whole island and we found another port in the middle...[where] there is a large community of Indians (Ortega 1970a:422-423).

Estevan Carbonel de Valenzuela, one of the pilots on Ortega's 1632 voyage, testified that:

> in one of these ports [on Isla Espíritu Santo], which we named Puerto Escondido because it is sheltered by very high cliffs, there is a community of Indians who came on board [our vessel]. There are many very large caves [there] where the Indians seek shelter in the rainy season, because in the dry season they sleep out in the open....This ranchería has as many as three hundred Indians (Carbonel 1970:353-354).

Fray Juan Cavallero Carranco, who accompanied Francisco de Lucenilla's 1668 pearling expedition, reported:

> [We] came in sight of a large island offshore [Espíritu Santo] where, according to reports, we hoped to find many Indians and many pearls. However, we did not find a single Indian on the entire island (Cavallero 1966:51).

In 1685 Isidro de Atondo y Antillón sailed to Isla Espíritu Santo to hunt pearls and hired some of the island's men as divers:

> We gave them to understand that they should go in the launch...and that for each fifty of the large ones [pearl oyster shells] I would give them a knife (quoted by Bolton 1936:210).

Atondo also noted that the island was inhabited by entire families:

The children of this island are fair and well featured, and so are the women. The latter wear grass skirts with which they cover their thighs, and the young women from fifteen to twenty [years of age] cover their breasts with the skins of sea birds (quoted by Bolton 1936:209).

Archaeological Record

Islas Espíritu Santo and Partida Sur were the first Gulf islands to be investigated, and they have been the scene of intensive archaeological research since the 1990s. In January 1883 Herman ten Kate spent four days exploring the west side of Isla Espíritu Santo for burial caves. He found five caves in three localities containing secondary burials with bones painted with red ochre (Kate 1884; 1977:57-61). A decade later Léon Diguet found two additional burials on the island (Diguet 1898:43; 1973). All belong to the funerary complex known today as Las Palmas (Fujita 2006:95; Massey 1966:47-49).

In 1930, archaeologist Frederick Rogers, travelling with ornithologist Griffing Bancroft's bird-collecting expedition, recorded a site with some 50 "stone circles" (probably corralitos), shells of many species, numerous hearths, and a "spear point" (Rogers 1930:19, map). In 1981 archaeologists Baudelina García Uranga and Jesús Mora conducted limited surveys there (Fujita and Poyatos de Paz 1998:69). From 1994 to 2008, Harumi Fujita directed extensive surveys and excavations on both islands, recording 127 sites, including the largest site complex in the entire Cape region (Fujita 2002:2-4; Fujita and Poyatos de Paz 1998:69).

Habitation sites consist of shell middens, rockshelters (Fig. 3.3), and open-air camps, some of which include hearths, corralitos (Fig. 3.4), cairns (Fig. 3.5), and rock clusters (Fig. 3.6). Other sites include burial caves, rock art, and trails. Artifacts include flaked stone tools (including more than 20 projectile points), lithic manufacturing debris, metates (Fig. 3.7), manos, mortars, and worked bone and shell. Faunal remains, indicative of diet, include more than 70 shellfish species, bones of at least twelve species of fish, and bones of sea turtles, sea lions, and dolphins. Late period deer bones were presumably brought from the peninsula, but those from early contexts could be from a deer population that lived there before rising sea level separated Isla Espíritu Santo/Partida Sur from the peninsula. The faunal record as a whole suggests that the island was inhabited year around (Fujita 2002, personal communication 2009; Fujita and Poyatos de Paz 1998).

Isla Espíritu Santo/Partida Sur is the only Gulf island with extensive chronological data, which consists of 179 radiocarbon dates from 40 sites. Apart from the enigmatic early dates from Covacha Babisuri (see below), these range from an uncalibrated date of 11,284 ± 121 years BP to recent, indicating that these islands were occupied from the initial Holocene to the historic period (Fujita 2002, 2006:85-86, personal communication 2009; Laylander 2006:2-3). At Covacha Babisuri, a stratified rockshelter, the upper levels produced a series of 26 calibrated and corrected radiocarbon dates (mostly on shells) spanning the Holocene from approximately 9000 BC to roughly AD 1460 (Fujita 2007:Tabla 1, personal communication 2009; Fujita and others 2006:Cuadro 1). In the lower levels, shells associated with lithic artifacts yielded a spectacular suite of 20 uncalibrated radiocarbon ages ranging from 36,550 ± 310 to greater than 47,500 years BP. The shells, however, may have been already ancient when collected by the site's inhabitants and hence predate the human occupation by thousands of years. The interpretation currently considered most likely is that lowered sea level in the late Pleistocene, between 21,000 and 10,000 BP, exposed previously inundated

Figure 3.3. Isla Espíritu Santo, southwestern side. General view of Covacha Babisuri, a rockshelter with a long series of radiocarbon dates indicating occupation throughout the Holocene. Photograph looks northwest. October 2003. Photograph courtesy of Harumi Fujita and Instituto Nacional de Antropología e Historia.

Figure 3.4. Isla Espíritu Santo, southwestern side. One of about 40 corralitos at a large open-air camp. Photograph looks northwest. October 2006. Photograph courtesy of Harumi Fujita and Instituto Nacional de Antropología e Historia.

Figure 3.5. Isla Espíritu Santo, southwestern side. Four small rock cairns along the edge of a cliff at an open-air camp. Photograph looks northwest. October 2006. Photograph courtesy of Harumi Fujita and Instituto Nacional de Antropología e Historia.

Figure 3.6. Isla Partida Sur, eastern side. Rock cluster (foreground) and two small rock cairns (center, near the edge of the cliff). These structures are part of a large open-air camp that includes two corralitos, several more cairns and clusters, several manos and metates (see Fig. 3.7), and debitage from the manufacture of flaked stone tools. Photograph looks northwest. December 2002.

Figure 3.7. Isla Partida Sur, eastern side. Metate in situ with the grinding surface exposed. It is one of several metates at the open-air camp shown in Figure 3.6. December 2002.

natural beds of 40,000-year-old shells, which were collected during that interval not for food, but as implements or containers (Fujita 2007, personal communication 2009; Fujita and others 2006:71).

Although many archaeological sites on Espíritu Santo/Partida Sur are undoubtedly assignable to the historic Pericú Indians and their prehistoric ancestors, determining the ethnic identity of specific remains may be challenging. Fujita and Poyatos de Paz (1998:90) speculated that one of the island's most important sites, La Ballena 3, which includes semi-submerged stone structures that are probably fish traps, was the Pericú settlement visited by Ortega in 1633. However, Herman ten Kate visited Bahía Ballena in 1883 and wrote:

> At the beach of Ballena Bay, there are some stone dikes, used for catching fish. They are filled with water at flood tide, but at ebb tide the water flows away slowly through the fissures between the stones of the dikes and leaves the fishes and mollusks as an easy prey for the Yaqui Indians, who made these dikes (Kate 1977:58-59).

If these are the same structures, they were indeed made by Indians, but these Indians would have been nineteenth century Yaqui pearl divers from Sonora living under Mexican influence. These latecomers might also have brought the small quantity of historical-period pottery found on the island.

ISLA SAN FRANCISCO

Water Resources

According to Frederick Rogers, one of the island's canyons has three tinajas "from which native fisherm[en] obtain water" (Rogers

1930:25, map). Griffing Bancroft added that:

> Here we had for company our friends the pearlers, who had come ashore for water....They were taking advantage of the fact that some of the rain [during the past three days] had accumulated in a string of small basins along the bottom of one of the ravines. The Mexicans had brought ashore six or eight empty gasoline cans which, with considerable labor, they had filled [with water] and had carried back to the beach (Bancroft 1932:207).

Sailors Leland Lewis and Peter Ebeling, old hands in the Gulf, noted that:

> Tinajas, pools of brackish water, may be found for some time after the rainy season in the arroyos and their existence is known and utilized by the natives (Lewis and Ebeling 1971:245).

Documentary Record

Francisco de Ortega, who visited Isla San Francisco in 1633, described it as:

> 6 leagues in circumference. On October 24 we dropped anchor on its western shore and, having just anchored, Indians came down from the hills saying, "Captain, friend—*boo*," which [is their word for] pearls. They came on board by swimming [to the ship] where they gave us many grains of very good pearls....these Indians are very friendly [toward the Indians who live] at the Port of La Paz, all of whom speak the same language (Ortega 1970b:440-441).

Writing in 1746, after the Isleños had been relocated to the peninsular missions, William Strafford noted that:

> there is an island named San José...and another little island named San Francisco

which are depopulated (Stratford 1958:55).

Archaeological Record

Isla San Francisco has never been formally surveyed. In 1930, Frederick Rogers observed a series of eight rockshelters containing shells and flaked stone (Rogers 1930:25, map). A brief shore visit in 2004 turned up a small site on the northern part of the island containing about 20 andesitic flakes, some retouched, and half of a small granitic beach pebble (Bowen:field notes).

ISLA SAN JOSÉ

Water Resources

Francisco de Ortega, who visited Bahía Amortajada at the southern tip of Isla San José in 1633, stated that:

> Along the beach inside this bay there are many very good springs of fresh water and a quantity of firewood (Ortega 1970b:441).

In 1746, William Strafford wrote that:

> in the said [island of] San José, there is water with which those who go to the pearl fisheries are provided (Stratford 1958:55).

In the 1870s, the crew of the U.S.S. *Narragansett* visited Bahía Amortajada and found that:

> Fresh water may be obtained here (Belden 1880:81).

Lewis and Ebeling reported that:

> Brackish water standing in pools may be found in the arroyos that open into the shore in the vicinity of Punta Colorado....

[and] fresh *water* rises at the head of Bahía Amortajada and at the settlement at Punta Salinas (Lewis and Ebeling 1971:247).

Documentary Record

In 1633, Francisco de Ortega described Bahía Amortajada at the southwestern tip of the island:

> [Isla San José] measures sixty leagues in circumference and is two leagues from the peninsula. There is a large port on its western shore with two reefs across its entrance. When we entered this port in the launch we found many Indians of the same language and character as those at the Port of La Paz. This bay has a circumference of four leagues and the local Indians took us to a shell midden inside it...where [our] divers and the Indians...obtained high quality pearls (Ortega 1970b:441).

Three years later Ortega:

> dropped anchor at the northern end of the west coast of Isla San José.... The local Indians showed us a shell midden which we found to be rich in pearl oyster shells....The local Indians gave us burned and drilled pearls which they said they had obtained from this shell midden (Ortega 1970c:461).

In 1683, Father Eusebio Francisco Kino commented on:

> the docile Indians of the Island of San José. It is evident to us that they already have some knowledge of things Catholic. Towards the end of [the previous] April when the [vessel] Capitana was on its way to Yaqui, the inhabitants of that island called out to our people, and reached the ship by swimming. As soon as their chieftain saw a picture of Our Lady of Guadalupe, he revered it and made the sign of the cross (Kino 1954:35).

In 1709, Father Juan María de Salvatierra, commenting on a recent incident, asserted that:

> Here we have proof that the Indians of [Isla] San José slew the unlucky [pearl] divers who came in a boat from Colima (Salvatierra 1971:225-226).

Archaeological Record

Although the island has never been formally surveyed, several sites have been reported, mostly in the vicinity of Bahía Amortajada. In 1930, Frederick Rogers recorded a series of shell middens, a shell scatter, a rockshelter containing shells, and a series of four fire pits lying within stone circles. Apparently, there was no clear way to separate indigenous from modern components at some of these sites (Bancroft 1932:203; Rogers 1930:map).

In 1957, biologists from the American Museum of Natural History investigated a series of large shell middens in the same vicinity and published a detailed report (Emerson 1960). Maximum depth of the deposits was 20 feet and the middens contained some charcoal but no evidence of bones or stone artifacts. Malacologist William Emerson identified 35 species of shells, all locally extant. A weathered *Lyropecten* shell from the upper portion of a midden yielded an uncalibrated radiocarbon date of 1000 ± 80 BP (Emerson 1960:5-8).

In the early 1990s, a salvage project on Isla San José led by physical anthropologist Alfonso Rosales López recovered a partial human burial (Harumi Fujita, personal communication 2008). A brief shore visit in 2002 revealed a tiny site on the western side of the island that includes a plane of silicious rock, a broken and flaked beach cobble, and a few shells (Bowen:field notes).

ISLA SANTA CRUZ

Water Resources   No data.

Documentary Record   None known.

Archaeological Record

The island has never been formally surveyed, and no archaeological remains are known.

ISLA SANTA CATALINA (CATALANA)

Water Resources

In 1952 or 1953, "A shed [snake] skin was found near a small spring" by herpetologist Frank Cliff (1954:75). George Lindsay wrote that during the 1962 Belvedere Expedition:

> The ship was moved to a rock-and-sand spit on the northeast side of Catalina. A canyon south of the spit contains a small spring and a grove of date palms, which were visited by [Reid] Moran, [John] Harbison, and [Michael] Soulé (Lindsay 1962:29).

In 1981, during a natural history cruise, Lindsay and others encountered a second spring on the eastern side of the island:

> [Naturalist guide] Peter Butz found a freshwater spring with a date palm and sedges, and many birds. He said it was not far and many made the hike. Peter is young and his legs are long, as was the walk to the spring. A fox was seen (Lindsay and Lindsay 1981:8).

Peter Butz recalled in 2008 that the spring was fairly high up a steep arroyo. He also remembered the putative fox sighting but did not recall any details (Peter Butz, personal communication, 2008).

According to Lewis and Ebeling:

> seasonal tinajas, natural rock basins filled with still water, and two small springs provide a limited source [of fresh water] on Isla Catalan; a few stunted palms grow near one of the springs and mark its existence (Lewis and Ebeling 1971:258).

Documentary Record

In 1633 Francisco de Ortega visited an island he named Santa Cruz, which may have been today's Isla Santa Catalina:

> [We] arrived at another island...which we named Santa Cruz [and] where we dropped anchor. We did not find people on it although there are mother-of-pearl shells; the mainland [that is, peninsular] Indians cross over to this island to fish. At the southern end we found a pearl-bearing shell midden (Ortega 1970b:442).

On November 1, 1720, Fathers Jaime Bravo and Juan de Ugarte set sail from Loreto bound for La Paz:

> Departing from the Bay of Loreto at 9 in the morning with a favorable wind, we passed the islands of Carmen, Monserrate, and Catalana. We wanted to go ashore on this island [Catalana] because some divers had told us that the Isleños from San José were there. The divers were given to understand that they [the Isleños] were going to Loreto to see the Fathers....But since the wind had greatly increased and the island had no port, we kept going. We supposed that the Isleños had returned to San José, because more than twenty days had passed since we encountered the divers, and the wind was entirely wrong [for them] to go up to Loreto (Bravo 1970:26-27).

Writing in 1772, Father Johann Baegert recalled an episode that took place early in the eighteenth century:

> Some of these islands were still inhabited in this century, especially the three: Catalána, Ceralbo, and the one called San José. Either by trading or by force, the inhabitants of these islands had acquired small boats from the pearl fishers and begun to practice piracy. However, after they were put out of trade in 1715 by [Baja] California soldiers, some of them died out, and others were transferred to the missions in [Baja] California. These and other islands are now deserted (Baegert 1952:12).

Archaeological Record

Although the island has never been formally surveyed, brief shore visits in 2003 and 2007 resulted in the discovery of two small sites. One, on the high sea cliffs above the eastern coast, consists of two stone circles (Fig. 3.8), a rock cluster, about ten flakes from two different cores, a hammerstone, an unworked beach cobble, several broken beach pebbles, about ten whole limpet shells, and fragments of other shells. The other site, located on the western shore, has been used by recent fishermen but a few flaked beach cobbles and a broken but well-used metate suggest earlier use by indigenous people (Bowen:field notes).

Figure 3.8. Isla Santa Catalina, eastern side. Stone circle of widely-spaced stones, approximately 1.5 m in diameter. It is part of a small site that includes a thin scatter of flakes and shells. Photograph looks south. January 2008.

Remarks

Ortega's geography between Islas San José and Carmen is garbled, perhaps beyond redemption. In 1683, fifty years after Ortega's voyage, Father Eusebio Kino mapped the region (Bolton 1936:109, opposite 160), using Ortega's diaries, which he had borrowed, to supplement his own observations (Michael Mathes, personal communication 2008). Kino ignored the tiny islet of San Diego, placed Islas Santa Cruz, Santa Catalina, and Monserrat in their correct positions, and then apparently tried to reconcile these islands with Ortega's diaries, using Ortega's names. Thus Kino labelled today's Isla Santa Cruz "San Diego" and today's Isla Santa Catalina "Santa Cruz." Ortega and Kino both bestowed the name "Monserrat" on the island that goes by that name today. The lingering question, then, is whether Kino's "Santa Cruz" —today's Isla Santa Catalina—is also Ortega's "Santa Cruz" and the island where Ortega found evidence of Indians, if not the people themselves. By the mid-1700s the Ortega/ Kino name "Santa Cruz" had been replaced by "Catalana" and other variants, and by 1829 the island had become "Santa Catalina" (Hardy 1977:map).

ISLA MONSERRAT

Water Resources

No specific resources are known. Historians Peter Gerhard and Howard Gulick speculated that:

> At one time Monserrate must have had fresh water, as it was inhabited by Indians as late as 1717 (Gerhard and Gulick 1970:207).

Documentary Record

In 1717 Father Jaime Bravo, in a letter to the Viceroy, recommended creation of a mission at La Paz to quell the hostility of the Guaycura Indians:

> I believe it would be useful to install a Father Missionary in the center of [Guaycura] territory, along with a squad of fifteen soldiers and a satisfactory Corporal....This mission can be established at the port of La Paz....With this [done], the Father Missionary will be able to go out and pacify the Guaicura nation, attracting to it [the mission] the gentiles of the islands of Espíritu Santo, St. Joseph, and Monserrate (Bravo 1979:209).

Archaeological Record

The island has never been formally surveyed, and no cultural remains are known.

ISLA DANZANTE

Water Resources   No data.

Documentary Record   None known.

Archaeological Record

Although the island has never been formally surveyed, cultural remains have been reported near the northern tip of the island:

> Evidence of Indian middens may be found in the arroyos back of these coves and beaches (Lewis and Ebeling 1971:264).

Remarks

The original name for Isla Danzante, bestowed on it by Ortega in 1633, was "Isla de las Pitahayas." Ortega named the bay on the peninsula directly opposite the island "Bahía de los

Danzantes" because "the Indians we found in that bay greeted us dancing and playing flutes made of canes" (Ortega 1970b:444). The name "Danzantes," however, was eventually transferred to the island and the bay was renamed Puerto Escondido.

ISLA CARMEN

Water Resources

In 1699, Father Juan María de Salvatierra wrote about the discovery of fresh water in Puerto Balandra on the western side of Isla Carmen:

> although the early explorers did not encounter fresh water on Isla Carmen, we have been given the good news that it has sources of fresh water, news that will be a relief and will facilitate navigation (Salvatierra 1946:94).

Several later visitors reported water on the eastern coast, probably at the source known today as Agua Grande. Apparently referring to this source, in the 1870s the crew of the U.S.S. *Narragansett* were told that south of Salinas:

> there is a small stream where it is said fresh water may be procured (Belden 1880:94).

Edward Nelson did not visit Isla Carmen during his 1905-1906 survey of the peninsula, but he too was told that:

> Fresh water is found here [at Bahía Salinas], and about halfway from Salinas Bay to the southern end of the island on the east coast is a small stream of fresh water (Nelson 1922:93).

Herpetologist Joseph Slevin wrote that the 1921 California Academy of Sciences expedition found water near:

Agua Grande, five miles south of Salinas Bay. About half a mile up the cañon from the anchorage is an excellent spring of fresh water (Slevin 1923:67).

Similarly, in 1962 George Lindsay noted that:

> A permanent spring supplies water to a large salt works at Salinas Bay (Lindsay 1962:21).

South of Salinas, Lindsay also noted that he and herpetologist Charles Shaw:

> went up a narrow arroyo in whose limestone bottom were several tinajas with damp sand and some water (Lindsay 1962:23).

Documentary Record

In 1633, Francisco de Ortega:

> went ashore on a very large island [which he named Isla del Carmen]...where we found Indians who approached us trembling. They are of a different nation and language from the others whom we have seen up to now on this trip (Ortega 1970b:444).

In 1699, Father Juan María de Salvatierra wrote to Father Juan de Ugarte about the discovery of Puerto Balandra:

> two vessels left from this bay [Loreto], retreating into a secure port on Isla Carmen which faces our Loreto, from which the canoes and balsas of the Indians come and go in mild weather (Salvatierra 1946:94).

Like the other islands, Carmen was abandoned by Indians during the early Jesuit period. Thus Father Miguel del Barco, who served on the peninsula from 1738 to 1768, wrote of Isla Carmen that "This island is uninhabited" (Barco 1980:284).

In the 1860s, Indians were back on Isla Carmen, but this time they were Yaquis from Sonora, brought over to work the island's salt deposits (Bowen 2000:120-121).

Archaeological Record

Although the island has never been formally surveyed, a few sites have been observed. In 1921, paleontologist Fred Baker collected shell specimens from "kitchen midden material" at both Marquer Bay and Puerto Balandra (Hanna and Hertlein 1927:144, 147). In 1930, Frederick Rogers saw large shell middens with "Much Chipped Stone and fire" at Puerto Balandra (Rogers 1930:31).

ISLA LOS CORONADOS

Water Resources

According to Peter Gerhard and Howard Gulick:

> There is no fresh water on the island (Gerhard and Gulick 1970:209).

Documentary Record   None known.

Archaeological Record

Although the island has never been formally surveyed, one site has been reported. In 1984, ornithology graduate student Luke George (personal communication 1984) found a site containing "several arrowheads" and "many rocks that had been worked lying nearby."

ISLA SAN ILDEFONSO

Water Resources

Francisco de Ortega, writing in 1636, stated that:

> On this island we found a water source in some pools where rain-

water collects, and the Indians have beach wells of brackish water that last all year (Ortega 1970c:462).

Entomologist Paul Arnaud noted that in March 1953:

> There was some fresh water caught in rock basins, and [John] Figg-Hoblyn collected two mosquitos. [Dallas] Hanna was bitten by mosquitos quite a few times (Arnaud 1970:15).

Documentary Record

In 1633, Francisco de Ortega reported that while:

> sailing North fourteen leagues from this island [Los Coronados], we found another island which we named San Ildefonso; it is four leagues from the mainland and twenty leagues in circumference; it is populated with people (Ortega 1970b:445).

Ortega returned three years later, arriving:

> at Isla de San Ildefonso on March 20 [1636]....the Indians we found on this island gave us some grains of burned and drilled pearls. [We] asked them where they obtained them, using signs [because] these Indians [speak] a language different from that at the Port of La Paz, which they do not understand. The Indians showed us a midden rich in pearl shells...in a cove on the Western shore.... The Indians showed us another midden of pearl [shells] on the North end of this island from which they obtain good pearls....These Indians are very warlike...[and] their weapons consist of the bow and arrow and hardwood throwing spears (Ortega 1970c:461-462).

Archaeological Record

The island has never been formally surveyed and few cultural remains have been reported. In 1930, Frederick Rogers found one piece of flaked stone and some obsidian on the island (Rogers 1930:33). Paul Arnaud noted that during his 1953 visit:

> Obsidian fragments were found over the lower southern slopes but no source could be found. It was not a pure material but was not incrusted [sic] or weathered. It may have been carried there by [Indian?] egg hunters (Arnaud 1970:15).

Remarks

Modern commentators have pointed out a serious problem with Ortega's estimate of Isla San Ildefonso's size:

> Ortega describes the island as being considerably larger than it is now.... Either there is some discrepancy through mistaken description in Ortega's original log, in its subsequent transcription and translation, or Isla San Ildefonso is sinking into the sea (Lewis and Ebeling 1971:282).

Transcription error is the most likely explanation because the surviving versions of Ortega's diaries are all copies that contain other errors (Michael Mathes, personal communication 2008). The figure attributed to Ortega for San Ildefonso's circumference is 20 leagues (about 84 km) whereas its actual circumference is on the order of 6 km. Because Ortega positioned the island approximately correctly with respect to nearby islands and the peninsular coast, and because there are no larger islands anywhere in the vicinity, the figure of 20 leagues must be an error. Perhaps an original figure of two leagues was mistranscribed as 20 leagues, in which case Ortega's size estimate would be off by less than a factor of two. That would be reasonable because Ortega overestimated the size of nearly every island he visited, often by much more than a factor of two. Presumably, not every island he surveyed is sinking into the sea.

It is no mystery why Indians would have been lured to such a small island because San Ildefonso is (or was) "one of the greatest bird islands in the Gulf" (Bancroft 1932:236; see also Arnaud 1970:15; Bancroft 1932:243; Banks 1963:52-53; Krutch 1968; Maillard 1923:452; Townsend 1923:5; Velarde and Anderson 1994:Table 1). With nesting seabirds providing a cornucopia of food, even a small quantity of water in tinajas and beach wells probably would have made Isla San Ildefonso irresistible to native people. It is not surprising that Ortega found Indians on this small island.

ISLA SAN MARCOS

Water Resources

In 1826, former Royal Navy Lieutenant turned pearl hunter Robert W.H. Hardy visited Isla San Marcos and, commenting on the feral goats there, remarked that:

> as there is fresh water at the northern extremity of this island, the animals have done well (Hardy 1977:278).

In the 1870s, the crew of the U.S.S. *Narragansett* found that:

> Fresh water may be obtained near the northern end of the island, and an abundance of goat's flesh may be had for the trouble of shooting the animals (Belden 1880:105).

Joseph Slevin wrote in 1921 that:

Water can be obtained by sinking wells, and at the south end, there are some probably permanent pools of water in the small cañons where there are a few groves of palms (Slevin 1923:61-62).

Frederick Rogers (1930:37) encountered an unspecified water source at the northern end of the island during his 1930 visit.

Documentary Record

Francisco de Ortega visited the island in 1636, naming it Isla de las Tortugas:

On October 22, we arrived at an island which is about twenty-five leagues from Isla San Ildefonso [and] three or four leagues from the peninsula, which we named Isla de las Tortugas. This island is thirty leagues in circumference [and]...the end which faces southeast has a good cove protected from the north, northwest, west, northeast, and east, with a good bottom for any ship. Within this cove we found a midden rich in pearl shells, [and] we named the port and the shell midden San Andrés. The Indians of this island did not wish to come to us (Ortega 1970c:462).

By the mid-eighteenth century, Indians no longer lived on the island but they still occasionally visited it. Writing in the 1770s, Father Miguel del Barco mentioned one such visit about 1765 during which the Indians discovered (or rediscovered) the island's large gypsum deposits:

When the Indians saw them, having gone to this unpopulated island in the canoe of the mission on some undetermined errand, they dug

some [gypsum] out and took it to their missionary (Barco 1980:283).

Archaeological Record

Although the island has never been formally surveyed, two sites are known. In 1930, Frederick Rogers found "much" flaked stone, shell, and obsidian, plus an obsidian "spear head" at the northern end of the island (Rogers 1930:37). A brief shore visit in 2007 turned up a small midden on the eastern shore containing a variety of shell species, animal bones including sea turtle, some burned, at least one metate and several possible manos, a few flakes, and a large quantity of broken local rock (Bowen:field notes).

ISLA TORTUGA

Water Resources

Geologist Carl Beal visited the island in 1921 and found it to be "devoid of fresh water" (Beal 1948:22). Gerhard and Gulick (1970:211) agreed that "There is no fresh water."

Documentary Record   None known.

Archaeological Record

The island has never been formally surveyed, and there are no known reports of cultural remains.

ISLA SAN PEDRO NOLASCO

Water Resources

Among scientists, the prevailing view long has been that Isla San Pedro Nolasco is waterless. In 1949, botanist Howard Gentry wrote that "No source of fresh water has been reported" (Gentry 1949:96). Some three decades later botanist Richard Felger and herpetologist Charles Lowe (1976:21) agreed that "There is no fresh water on the island." However, modern fishermen and local researchers have known for some time that the island has a small but permanent seep, which they refer to as Agua Amargo:

> This tiny freshwater seep emerges from high cliffs 5 or 6 m above high tide line on the east-central side of the island. This is the only freshwater source on the island and has sustained fishermen and others in trying times of misfortune....The water is highly alkaline, as indicated by the name *amargo* "bitter." It supports the only wetland plant, *Cyperus elegans*, on the island (Felger and Wilder 2009).

Documentary Record　　　None known.

Oral History

The traditional Seri name for Isla San Pedro Nolasco is *Hast Heepni It Iihom*, 'Rock Where the Iguanas Are' (Moser and Marlett 2005:350), a reference to the island's abundant endemic iguana *Ctenosaurus nolascensis*. Although the name suggests familiarity with the island and its fauna, apparently there is no specific oral history about the island (Felger and Wilder 2009; Nabhan 2003:88).

Archaeological Record

The island has never been formally surveyed, and there are no known reports of archaeological remains.

Remarks

Felger and Lowe (1976:21) stated that the island "has never been inhabited by man." However, Indians may have visited the island from time to time to exploit its resources. Herpetologists

have speculated that ancestors of the Seris collected iguanas from San Pedro Nolasco and introduced them on Isla San Esteban (Grismer 2002:120; Nabhan 2002:553; 2003:78, 88). If so, this must have taken place a long time ago as the two populations are currently considered separate species.

ISLA SAN PEDRO MÁRTIR

Water Resources

In 1976 Felger and Lowe stated that "There is no fresh water on the island" (1976:22). That still seems to be the consensus (Daniel Anderson, Richard Felger, Bernie Tershy, Benjamin Wilder, personal communications 2008).

Documentary Record

In the 1880s, Yaqui Indians were brought to Isla San Pedro Mártir from Sonora to mine guano. Ornithologist Nathaniel Goss, who visited the island in 1888, provided a few details:

> The Company has a large force of Yaquie Indians collecting the guano.... One hundred and thirty-five Indians were on the payroll and many had their families with them (Goss 1888:240).

Oral History

The traditional Seri name for Isla San Pedro Mártir is *Iicj Icóoz*, which has no other known meaning (Moser and Marlett 2005:401). Modern Sonoran fishermen say that during the Porfirian era, from the 1880s to about 1910, the island served as a sort of Gulag for political prisoners, many of them Indians, presumably Yaquis. Though unlikely, they think the prisoners were harassed by Seris, who would paddle out to the island on balsas to rob them (Bowen 2000:135).

ᅟ

Archaeological Record

Isla San Pedro Mártir has never been formally surveyed, although a day on the island in 1984 provided a brief glimpse of the historical remains from the guano mining era (Bowen: field notes). These include a great many stone walls and terraces built on the island's cliffs and lower slopes and at least one largely intact village (Bowen 2000:135-137). Goss' remarks indicate that many or most of these structures were built by Yaquis—hence indigenous people—but living and working under non-native conditions.

Remarks

Despite its isolation, Isla San Pedro Mártir has three major resources that would have made it attractive to native people—sea lions around the shoreline, nesting seabirds on the lower slopes, and fruit from the dense forest of dwarf cardon cacti on the summit plateau. Although the sites on the shoreline and lower slopes are probably long gone due to wave action and guano mining, it would be well worth surveying the relatively undisturbed summit plateau for archaeological evidence of native visitors.

ISLA ALCATRAZ (PELÍCANO, SAN JUAN BAUTISTA)

Water Resources

Felger and Lowe (1976:24) stated that "There is no fresh water on the island."

Documentary Record

Father Ignaz Pfefferkorn, writing in the 1790s of events in 1750, described Isla Alcatraz (editorially misidentified as "Tiburón") as a Seri refuge from Spanish military harassment:

> So they departed at night in utter silence with bag and baggage, women and children, and happily reached the shore of the Gulf of California. Here they always kept in readiness a number of small boats, used for fishing and made of wood and thick reeds. Boarding these vessels they made their way to a small island... about a mile and a half distant from the coast where they no longer needed fear Spanish arms (Pfefferkorn 1989:154).

In 1764, Father Juan Nentvig, also writing of Seri sanctuaries during this period, referred to Alcatraz by its Colonial name and slightly misplaced it:

> Other Seri shelters were the Tiburón Island...and San Juan Bautista Island, nine leagues southeast of the former and two leagues from the coast (Nentvig 1980:78).

Ethnographer WJ McGee (1898:190-191) noted that Seris hunted pelicans on Alcatraz during the late nineteenth century. In 1934, ethnographer-collector Edward H. Davis participated in a Seri pelican hunt on the island (Davis 1934:139-150; 1965:213-218). Felger and Lowe alluded to Seris visiting, but not living on, Isla Alcatraz:

> although it has been frequently visited, it is not known to have been occupied by man (Felger and Lowe 1976:24).

Oral History

The traditional Seri name for Isla Alcatraz is *Soosni*, which has no other known meaning (Moser and Marlett 2005:561). Seri oral history and Mexican documentary sources both recount the 1850 kidnapping of a young Mexican girl, Dolores (Lola) Casanova, by Seri leader Coyote Iguana (Bowen 2000:237-239). In the Seri version, she was initially taken to a rockshelter on Isla Alcatraz (Lowell 1970:148-153).

Archaeological Record

In 1921, Fred Baker collected shells on Isla Alcatraz from "kitchen midden material possibly" (Hanna and Hertlein 1927:150). In 1984 a prominent rockshelter, possibly the one where Coyote Iguana took Lola Casanova, still contained surface shells (Bowen: field notes). A half-day systematic survey in 2005 covered about 40 percent of the island, including essentially all of its level ground. The sole site recorded at that time consists of 18 sherds of historic period Seri pottery (Fig. 3.9) and a thin scatter of shells (Bowen 2005b).

Remarks

Illegal artifact collectors have been looting Isla Alcatraz for decades, which no doubt explains the scarcity of archaeological remains.

ISLA TIBURÓN

Water Resources

The Seris knew some 43 water sources on Isla Tiburón. About a dozen of these, including springs, tinajas (Fig. 2.2), and beach wells, were considered permanent (Felger and Moser 1985:81-84).

Members of the Cardona pearl hunting expedition were probably the first Europeans to see water on Tiburón when they anchored there in 1615 and spent "three days taking on water and firewood" (Cardona 1974:102). Since then, Tiburón's water has been extensively documented in the historical literature (for example, Hardy 1977:286-287; McGee 2000:85, 91-93; Ugarte 1958:24, 45).

Documentary Record

The earliest known record of Seris on Isla Tiburón dates from Nicolás de Cardona's 1615 visit:

Figure 3.9. Isla Alcatraz, central part. Two sherds of historic Seri pottery, left center and upper right of photograph. The sherd at the upper right is part of the rim of a large olla. April 2005.

I sailed onward [up the peninsular coast]...and saw the opposite shore...at a distance of eight or ten leagues to the east. I crossed to reconnoiter it and anchored there. I found that it was a large island inhabited by Indian fishermen who were naked, and whose women wear buckskin aprons made from deer skins and glass beads on their throats and ears. Desirous of knowing with whom they communicated, Christians or enemy, we asked them questions but we could not find out anything definite....This island was one league from the mainland and Indians from there came out to it, for after being there for three days taking on water and firewood many people had gathered there (Cardona 1974:102).

The documentary record of Seris on Isla Tiburón since the Cardona expedition is voluminous. For a broad selection of examples, see Bowen (2000), Felger and Moser (1985), Sheridan (1999).

Oral History

The Seri name for Isla Tiburón is *Tahéjöc*, which was apparently extended to the whole island from a camp of the same name. An archaic name for the island is *Hant Hamoíij Quiimt* 'Open Circle' (Moser and Marlett 2005:329,566). The Seris regard Isla Tiburón as their traditional homeland, and their own history, as they perceive it, is inextricably tied to the island. To cite one example, in the 1960s the Seris recalled from memory the names and locations of 95 traditional camps there (Moser ca. 1966).

Archaeological Record

In 1976 a Seri woman, upon hearing of an upcoming archaeological survey of Tiburón, quipped to Edward Moser, "They shouldn't have any problem—the whole island is a site!" Indeed, 62 days of field work at four localities carried out between 1967 and 1986 covered only a small fraction of the island but produced more than 100 sites (Bowen: field notes).

Many sites on Isla Tiburón are habitation sites and include open-air camps, rock shelters, and shell middens (Fig. 3.10). They typically contain substantial quantities of artifacts, shells, animal bones, and charcoal. Shell middens extend up to 1 km in length. One contains at least 2 m of stratified deposits and possibly as many as 6 m. Some quarry-workshop sites have tens of thousands of waste flakes and spent cores. Sites of unknown function include rocky slopes with hundreds of talus pits, hills with small corralitos but no artifacts (Fig. 3.11), long parallel lines of rocks, and concentrations of rock clusters, cairns, and ground figures (for example, Bowen 1976:Figures 21, 24, 27-30). Other structures include stone circles and pit ovens for baking agave hearts (Fig. 3.12). Many burials have been observed.

Pottery is abundant on Tiburón, spanning the typological (and presumably chronological) range of the Seri ceramic tradition, and includes sherd disks and fired clay figurines. Intrusive Trincheras pottery from northern Sonora and Yuman pottery from the lower Colorado River area suggest prehistoric trade. Metates and manos of sometimes carefully selected beach rocks are exceptionally numerous at some habitation sites. Flaked stone tools range from little more than broken rocks to well-crafted knives and projectile points. One manufacturing site produced several large lanceolate point fragments (Fig. 3.13) suggesting Paleoindian occupation before Tiburón became separated from the mainland by rising sea level. A poorly collected radiocarbon sample from a shell midden, composed of several unrelated human bones, gave an uncalibrated age of 1100 ± 300 radiocarbon years BP (Haynes and others 1966:20).

Figre 3.10. Isla Tiburón, northern coast. Archaeology in the making -- a Seri camp at Tecomate in 1946. The brush houses stand on the eastern end of a huge stratified shell midden. Tecomate was a Seri summer camp that was probably occupied more or less continuously from prehistoric times to the middle of the twentieth century. Photograph by William Neil Smith II, summer 1946. Courtesy of the University of Arizona Library Department of Special Collections (MS 316, Box 27, Folder 27/5, Color No. 2).

Figure 3.11. Isla Tiburón, northern coast. One of about 10 corralitos on and around a low hill. They range from about 1 m to 2 m in diameter; this is one of the larger ones. There are no accompanying artifacts. Photograph looks north, with the Sonoran mainland in the distance. January 1977.

Figure 3.12. Isla Tiburón, southwestern coast. One of several eroded pit ovens in the local area, almost certainly used for baking agave hearts. Rim crest is 4.8 m in diameter and stands 20 cm above present ground surface. Present depth of the pit is 40 cm. Much charcoal is strewn about the structure. These pits probably date from the late nineteenth and early twentieth centuries when Seris sometimes brought agave hearts from Isla San Esteban to Isla Tiburón for baking. January 1983.

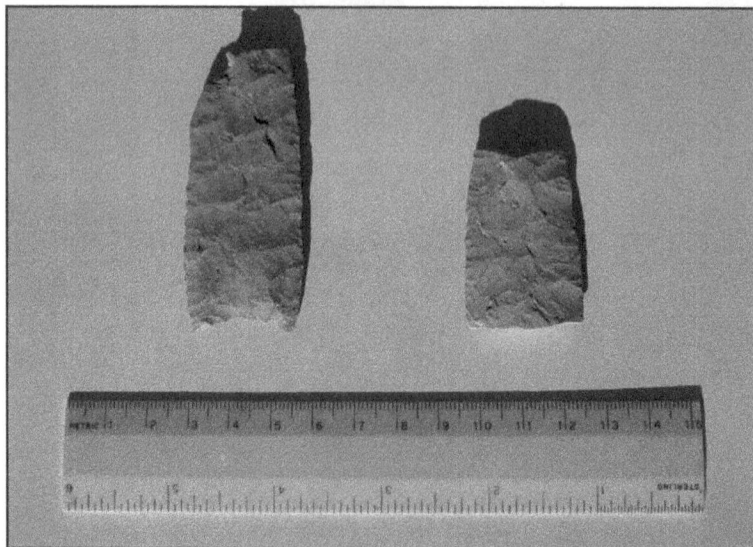

Figure 3.13. Isla Tiburón, northern coast. Two of several possible Paleoindian projectile point fragments of local rhyolite from a manufacturing site. The specimen at left, missing the tip and forward part of the left edge, is 6.7 cm long, 2.9 cm wide, and 5 mm thick. The base was thinned on both faces. The specimen at right is a midsection 4.3 cm long, 2.9 cm wide, and 6 mm thick. Basal fragments of two other points from the same site are closely similar in size, form, and flaking technique. The two points illustrated were stolen before they could be adequately documented. January 1976.

Not all Indians on Isla Tiburón have been Seris. In 1750 a Spanish military expedition spent two weeks on the island with a fighting force that included several hundred Upper Pima auxiliaries armed with bows and arrows (Pimentel 1999). In 1904, a small group of Yaquis who had raided a Mexican ranch took refuge on Isla Tiburón. Soon thereafter, a large Mexican military expedition which included 42 Tohono O'odham (Papagos) invaded the island, leaving a week later after the Yaquis had been killed (Moser 1988). Hence one should not automatically attribute cultural material —especially projectile points and skeletal remains—to the Seris or their prehistoric predecessors.

Remarks

Illegal artifact collectors have heavily looted some parts of the island.

ISLA DÁTIL (TURNERS)

Water Resources

According to Richard Felger (personal communication 2008), no fresh water has been found on the island.

Documentary Record

In 1850, a military expedition attempting to rescue Lola Casanova, a Mexican girl held captive by Seris, cornered and killed a small group of Seris on Isla Dátil (Bowen 2000:239).

Oral History

The traditional Seri name for Isla Dátil is *Hastáacoj* 'Large Mountain' (Moser and Marlett 2005:353). The Seris sometimes hunted seabirds on the island (Felger and Moser 1985:50).

Archaeological Record

The island has never been formally surveyed. A brief lunch stop on the island in 2000 revealed several clearings (Fig. 3.14), more than a dozen well-used manos, a sherd of Trincheras pottery from northern Sonora, and many shell fragments (Bowen:field notes).

ISLA PATOS

Water Resources

Felger and Lowe (1976:25) reported that "There is no fresh water."

Documentary Record

In 1826, Seris on Isla Tiburón indicated to Robert Hardy that they were familiar with Isla Patos: "The Indians say that quantities of seals [California sea lions] frequent this island" (Hardy 1977:296).

According to WJ McGee (1898:191), Seris hunted ducks on Isla Patos.

Oral History

The traditional Seri name for Isla Patos is *Hast Otíipa* 'Otíipa Mountain' or alternatively, *Hast Atíipa* 'Atíipa Mountain' (Moser and Marlett 2005:352). The Seris say that they formerly hunted sea birds and gathered eggs on Isla Patos (Felger and Moser 1985:50). They also recall their role as hired hands when the first ship came to mine guano on Patos in 1858 (Bowen 2000:123-124).

Archaeological Record

The island has never been formally surveyed.

Figure 3.14. Isla Dátil, eastern side. One of several circular clearings. A fragment of a white granitic mano is visible at the right edge of the clearing, and a similar mano lies nearby. Photograph looks north, with Isla Tiburón in the distance. March 2000.

No indigenous archaeological remains are known.

Remarks

Between the sporadic episodes of guano mining, beginning in 1858, and stripping the island of its vegetative cover in 1945 to encourage nesting (Bowen 2000:122-124, 138), it is unlikely that any archaeological record of indigenous people has survived.

ISLA SAN ESTEBAN

Water Resources

Seri oral history recalls five water sources. One was considered permanent even though it apparently failed periodically, forcing the island residents to retreat to Isla Tiburón (Bowen 2000:15-16). Archaeological survey has disclosed about 15 large tinajas and several smaller ones (waterless when recorded but with water stains visible on most). One set of five has a potential capacity of around 6000 liters (Bowen 2000:409-410; Villalpando 1989:30, 62).

Documentary Record

No unambiguous historical accounts of indigenous people are known.

Oral History

The traditional Seri name for Isla San Esteban is *Coftéecöl* (formerly spelled Coftécöl), or sometimes *Costéecöl* (Moser and Marlett 2005:204). Although previously glossed 'Chuckwallas-large' for the island's large endemic chuckwalla *Sauromalus varius* (Bowen 2000:5; Felger and Moser 1985:132), it is now thought that the name refers to the San Juanico tree (*Jacquinia pungens*), which

is said to make the same rustling sound as the chuckwallas (Felger and Moser 1985:372; Stephen Marlett, personal communication 2003; Moser and Marlett 2005: 204;). The modern Seris say that San Esteban was inhabited by a small band of Seris who were distinct in culture and dialect. They were sometimes called *Xica Hast Ano Coii* 'They Who Live in the Mountains' because Isla San Esteban is so rugged (Bowen 2000:5; Moser 1963:16). Much is remembered about these strange folk, whom the other Seris regarded as unsophisticated "country cousins." Their staple plant food was the endemic agave (*Agave cerulata dentiens*), and they also ate chuckwallas, fish, sea lions, and sea turtles. Although they were experts in the use of the balsa and travelled to Isla San Lorenzo and Baja California, they knew little of the outside world and were unfamiliar with soldiers and firearms. Consequently, they were easily rounded up and exterminated by a Mexican military expedition in the late nineteenth century (Bowen 2000:5-30; Moser 1963).

Archaeological Record

Extensive surveys totalling about 83 days, mostly conducted between 1980 and 1987, have covered about 35 percent of this rugged island's surface, including most of its accessible terrain. Results of these surveys have been published (Bowen 2000:315-394; Villalpando 1989). Excavation of a shallow rockshelter in 1984 by Elisa Villalpando and Arturo Oliveros remains unpublished.

The Seris say the San Esteban people had few material goods, and this is born out by the archaeological record. Habitation sites are rockshelters (Fig. 3.15) and open-air camps with oval clearings. They typically contain a few flakes, a metate or mano, a few shells and animal bones, sometimes pottery, and occasionally charcoal from a hearth. A few small quarry-workshop sites are known, and some of the rocky slopes have dozens to hundreds of talus pits. Clusters of corralitos and piled rock cairns of unknown function are perched on several high mountain summits (Fig. 3.16).

Most cultural remains occur as isolated and dispersed structures and artifacts. Structures include rock clusters and cairns, but the majority are stone circles (Fig. 3.17). Unlike those on other islands, most circles are associated with charcoal-laden soil and are thought to be areas where agave hearts were baked, using the peninsular surface-baking method. Agave knives, usually tabular slabs with a few flakes hastily removed from one edge, are the only common flaked stone tools, and about 200 have been found (Fig. 3.18). Other flaked artifacts include scrapers, core choppers, and oval bifaces. Metates and manos, some well used, are common. Pottery, comparatively scarce, is exclusively of Seri manufacture. Although it is said that the San Esteban people lacked the bow and arrow, ten projectile points are known; all are small leaf-shaped bifaces or small triangular points with concave bases (Fig. 3.19). The former seem to have little chronological value in the region, while the latter are widely known from late prehistoric and historical contexts (Hyland 1997:298-300). Four calibrated radiocarbon samples from surface charcoal are also late, all postdating AD 1650 (Bowen 2000:317-394).

Remarks

The archaeological record is consistent with the broad outlines of Seri oral history, indicating that a small band of Seris resided on Isla San Esteban until the late nineteenth century.

ISLA SAN LORENZO

Water Resources

In 1972, geology graduate student Robert Ros-siter (personal communication 1983) found a tinaja containing water. Its maximum capacity is on the order of 500 liters. A second possible tinaja has a capacity of about 1650 liters but has not been observed with water (Bowen 2005a:402; field notes).

Documentary Record

In 1802, after an Indian attack on Santa Gertrudis, a peninsular mission about 60 km southwest of San Lorenzo, Interim Governor José Joaquín de Arrillaga speculated that the perpetrators were Seris who used (unspecified) islands of the San Lorenzo archipelago as a staging area (Arrillaga 1999:452-454). WJ McGee (1898:49) claimed that Seris travelled to San Lorenzo for a mineral pigment for face painting.

Oral History

The traditional Seri name for Isla San Lorenzo is *Coof Coopol It Iihom,* 'Where the Black Chuckwalla Is' (Moser and Marlett 2005:216). The Seris say that the San Esteban people frequently visited San Lorenzo to gather an especially flavorful variety of wild tobacco (*Nicotiana obtusifolia*) (Felger and Moser 1985:369). They also went there when food ran low, staying up to a year and subsisting on the abundant black chuckwallas (*Sauromalus hispidus*), sea lions, and fish. Using genealogies, Edward Moser estimated that some of these trips took place between 1820 and 1835 (Bowen 2000:23-24, 471-472).

Expeditions of Tiburón and mainland Seris to San Lorenzo were evidently rare. It is said that they sometimes quarried a special clay on visits to San Lorenzo for use in their pottery (Bowen and Moser 1968:92). It is also possible that the black chuckwallas introduced on Isla Alcatraz were brought there by Seris from Isla San Lorenzo or its neighbor islands (Grismer 1994:30; Nabhan 2002:412).

Figure 3.15. Isla San Esteban, eastern side. Partly collapsed rockshelter in the bank of a small arroyo. Surface artifacts and ecofacts are sparse but varied, including two metates, a mano, a chopper, a scraper, two agave knives, a sherd of historic Seri pottery, plus four shells, several fragments of sea turtle plastron bone, and a small quantity of charcoal. A charcoal sample submitted for radiocarbon analysis proved to be contaminated and hence undatable. January 1981.

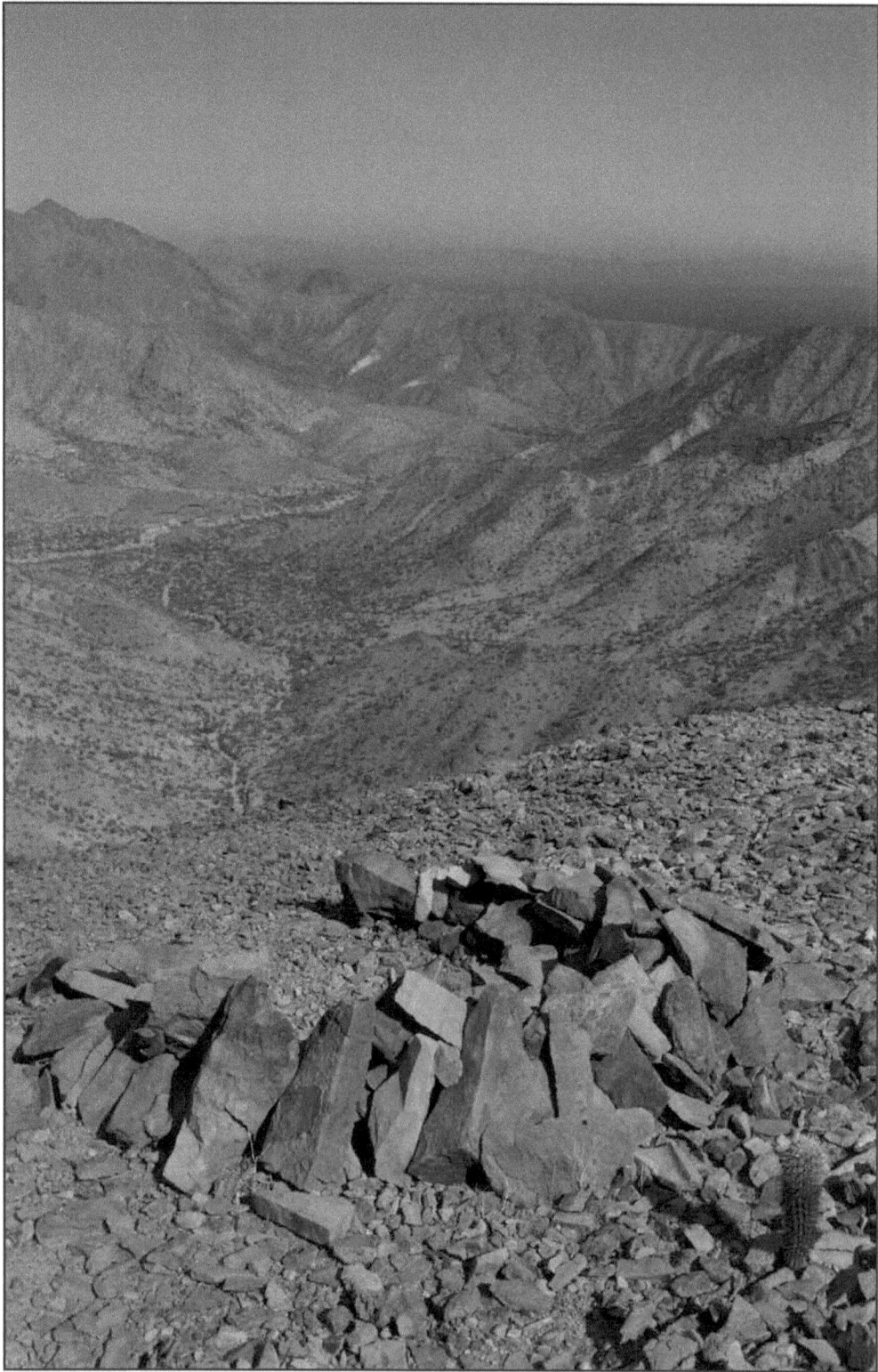

Figure 3.16. Isla San Esteban, central part. One of several corralitos on the summit of a prominent peak. Length is 1.5 m, width 1.0 m (inside dimensions), maximum height 60 cm. The opening, 70 cm wide, faces north. Most of the island's high summits have corralitos but no artifacts, suggesting that these summit structures were lookout posts of some kind rather than residential structures. Photograph looks north-northeast. January 1982.

Figure 3.17. Isla San Esteban, eastern side. One of more than 60 stone circles recorded on the island. Inside diameter is 1.9 m. Like most San Esteban circles, this one was built on a patch of charcoal-laden soil, possibly to mark the location where agave hearts were baked. As the photo indicates, agave is abundant in the immediate area, and two agave knives were found nearby. Photograph looks east-southeast. January 1987.

Figure 3.18. Isla San Esteban. Bifacially-flaked agave knife of local tabular andesite, in situ. Length is 12.5 cm, width 8.5 cm, thickness 4.5 cm. December 1980.

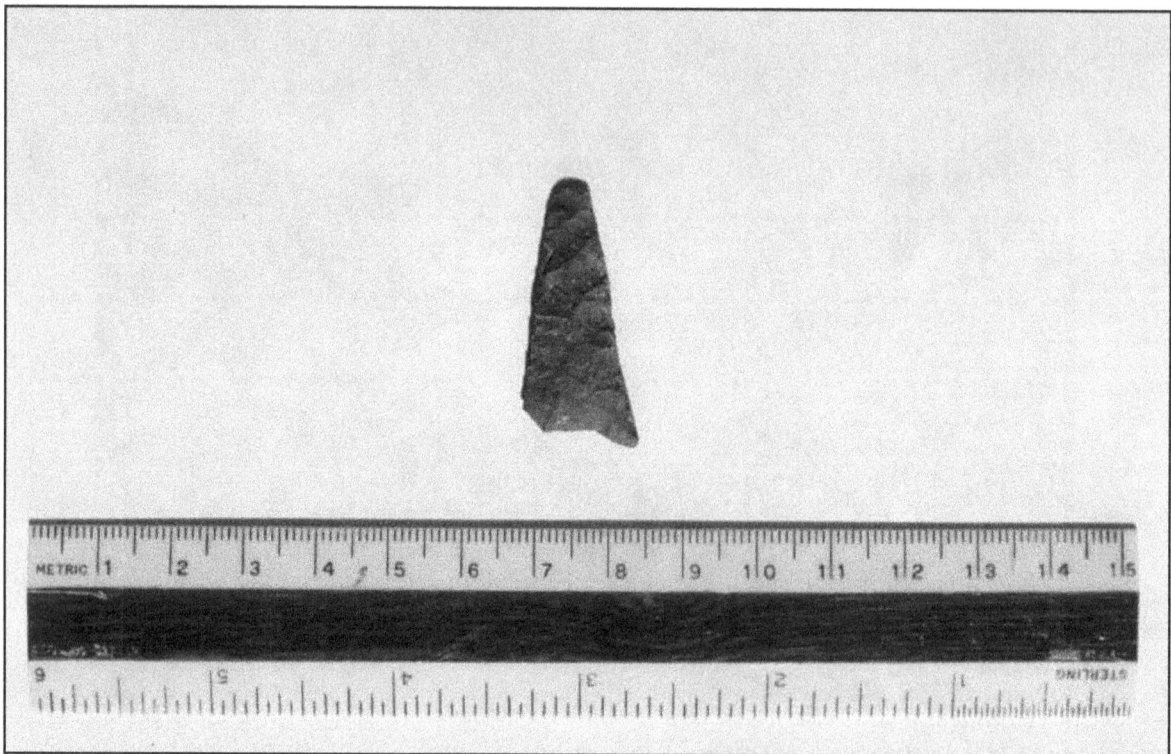

Figure 3.19. Isla San Esteban, northeastern corner. Small triangular concave-base projectile point of rhyolite, with a broken left tang and tip, from the surface of a small rockshelter. Length is 3.5 cm, width 1.5 cm, thickness 3 mm. April 1980.

Archaeological Record

Extensive surveys on Isla San Lorenzo totalling some 25 days, conducted mostly in 1984 and 1985 and again between 2005 and 2007, have covered about 10 percent of this precipitous island, including perhaps half of its accessible terrain. Cultural material is scarce and limited mostly to a few small sites. These include a small camp, two quarry-workshop sites (Fig. 3.20), three small concentrations of flakes and tools (Fig. 3.21), and a site of unknown function consisting of a set of rock alignments (Fig. 3.22) and three stone circles, one with outlier stones accurately marking the cardinal directions (Fig. 3.23). The most noteworthy site is a Seri camp, identified by its distinctive pottery. In addition to pottery, artifacts at this site include metates and manos, flaked stone choppers and denticulates, and spent cores and waste flakes. Interestingly, there are 17 flaked *Dosinia* shells, some of them heavily used (Fig. 3.24), along with debitage from their manufacture. The pottery, from three different vessels suggests repeated visits during the late eighteenth century and during the nineteenth century, and confirms Seri oral tradition of trips to the island (Bowen 2005a).

Figure 3.20. Isla San Lorenzo, central part. Small quarry-workshop site at about 300 m elevation on the central spine of the island. The site covers the flats where the two people are standing. The two white patches behind the people are outcrops of quartz, the main quarry rock, but a small amount of local vitreous black basalt was also exploited. The site includes a small stone circle and a linear alignment of about 16 rocks with rock clusters at each end. Photograph looks southeast. April 2005.

Figure 3.21. Isla San Lorenzo. Double denticulate made on a primary flake of local rock, flaked unifacially and retouched to shape the teeth. Like most denticulates, the projections are accentuated by concavities on either side, and these concave edges may have been used for scraping. The prominent flake struck from the edge opposite the projections was clearly intentional and forms a perfectly-positioned thumb rest. Length is 6.8 cm, width 5.9 cm, thickness 2.2 cm. April 2005.

Figure 3.22. Isla San Lorenzo, central part. Rock alignment, one of six rock figures on the surface of a small playa at about 300 m elevation on the central spine of the island. It is a V-shaped alignment with arms of unequal length. The longer arm, about 18 m in length, curves inward and is divided into two segments by a short gap in the rocks. The shorter arm, 5.4 m long, is straight and continuous. The "V" forms an angle of about 45 degrees and points almost due west. Photograph looks northeast. January 1984.

Figure 3.23. Isla San Lorenzo, central part. Stone circle, 2.9 m in diameter. It is one of six figures on the surface of a small playa at 300 m elevation on the central spine of the island (see Fig. 3.22). It consists of small stones placed between eight sets of larger stones. Five of the sets of larger stones are accompanied by additional large stones radiating outward from the edge of the circle. The large stones apparently mark the cardinal compass points, and the north-south axis is accurate within three degrees of true. Photograph looks east. January 1984.

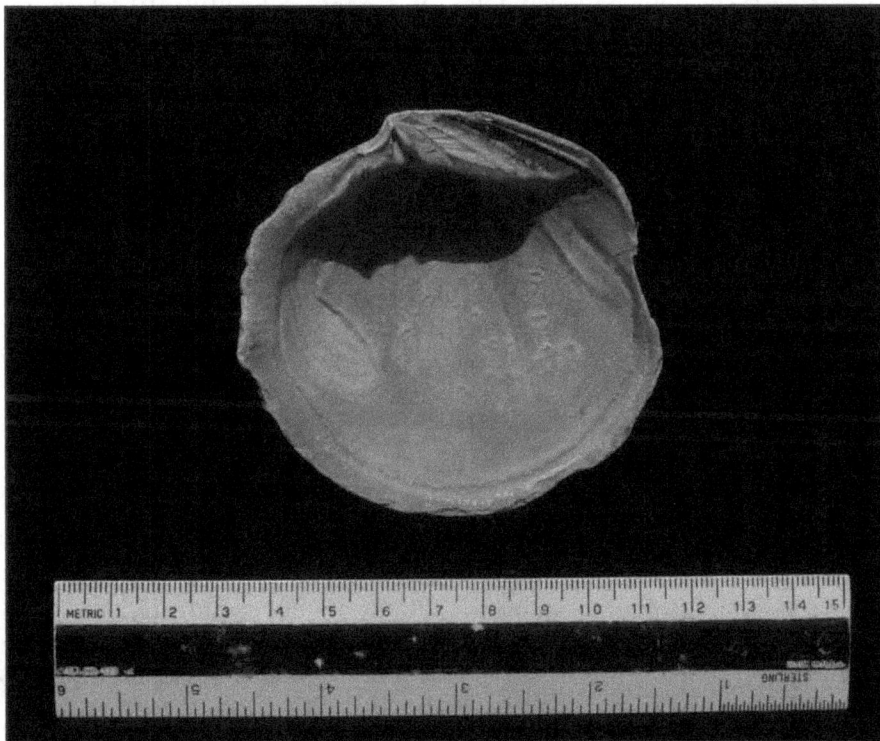

Figure 3.24. Isla San Lorenzo, western coast. Flaked shell of Dosinia ponderosa *at the Seri camp. The shell, flaked and retouched from the opposite (convex) side, was so heavily used that the edge was worn smooth and the flake scars largely obliterated. Diameter is approximately 8.5 cm. March 2004.*

Isla Las Ánimas (San Lorenzo Norte)

Water Resources

In 1984, avian ecologist Daniel Anderson (personal communication 1984) observed several large tinajas filled with water after a hard July rain.

Documentary Record   None known.

Oral History

The traditional Seri name for Isla Las Ánimas derives from their name for the narrow channel that separates it from Isla San Lorenzo. The strait is called *Hant Iicot Conttaca Toii Hant Cöicáap* 'The Place One Passes through to Go toward Another Place' or, more freely translated, 'Strait to the Land Beyond.' The "Land Beyond" is Baja California, and the Seris viewed the strait as a shortcut between Isla Tiburón and the peninsula. The name for Isla Las Ánimas itself is *Hant Iicot Conttaca Toii Hant Cöicáap Hast* 'Mountain at the Place One Passes through to Go toward Another Place' (Stephen Marlett, personal communication 2004; Moser and Marlett 2005:330). Apparently, the Seris have no specific recollection of early visits to Las Ánimas, as distinct from Isla San Lorenzo.

Archaeological Record

Extensive surveys totalling about nine days, mostly conducted in 2004 and 2005, have covered about 30 percent of the island, including about half of its accessible terrain. The archaeological record is substantial but dispersed. Eight small flake scatters or quarry-workshops have been recorded, one of which includes two stone circles. Most remains are individual structures and artifacts scattered throughout the island. Structures include at least 11 full or partial stone circles and three rock clusters (Fig. 3.25). Artifacts consist of simple flake and core tools (Fig. 3.26) and manufacturing debris. There are no obvious camps, no metates, and only a single questionable mano (Bowen:field notes).

Remarks

Although Islas Las Ánimas and San Lorenzo are separated by a very narrow channel, unlike Islas Espíritu Santo and Partida Sur, these two islands do not constitute a biological unit.

Isla Salsipuedes

Water Resources   None known.

Documentary Record

WJ McGee (1898:49) claimed that Seris occasionally visited the island but offered no details.

Oral History

The Seri name for Isla Salsipuedes is *Tatcö Cmasol It Iihom* 'Where the Yellow Leopard Groupers Are' (Moser and Marlett 2005:567).

Archaeological Record

Extensive surveys totalling about six days, conducted mostly in 2004 and 2005, have covered about 80 percent of the island, including nearly all of its accessible terrain. Cultural material recorded includes five stone circles (Fig. 3.27), two partial circles, a rock cairn, two rock clusters, a small number of flakes, cores, and possible tools (Fig. 3.28), a well-used metate, and a few *Turbo* shells (Bowen:field notes).

Figure 3.25. Isla Las Ánimas, north end. Rock cluster, 1.3 m in diameter, consisting of about 30 rocks. Photograph looks east. February 2004.

Figure 3.26. Isla Las Ánimas. Domed bifacially-flaked scraper of white quartzite. The cortex was completely removed from both faces except for the small patch in the center of the exposed face. Length is 8.3 cm, width 4.8 cm, thickness 2.8 cm. February 2004.

Figure 3.27. Isla Salsipuedes, northern end. Stone circle with widely-spaced stones set 30 cm to 50 cm apart. It is 2.0 m long and 1.7 m wide, inside dimensions. Scattered rocks 3 m away might be the remnants of another circle. Photograph looks east. February 2004.

Figure 3.28. Isla Salsipuedes. Bifacial chopper of local volcanic rock. The working edge was created by detaching three flakes from the exposed face and one from the opposite face. The edge has been dulled from use. Length is 9.7 cm, width 6.2 cm, thickness 3.1 cm. February 2004.

Isla Rasa

Water Resources   None known.

Documentary Record

WJ McGee (1898:49) implied that Seris may occasionally have visited Rasa, but presented no evidence.

Oral History

Several fanciful theories have tried to link the Seris with Isla Rasa (Bowen 2000:129-130). These aside, the traditional Seri name for the island is *Tosni Iti Ihíiquet* 'Where the Pelicans Have Their Offspring' (Moser and Marlett 2005:575). Apparently, however, the Seris have no specific recollection of early visits to Rasa.

Archaeological Record

The island has never been formally surveyed, although it has been possible to record some of the island's historical remains during brief visits, totalling about two days, in 1984, 1993, and 2004 (Bowen: field notes). Guano mining on an industrial scale began in 1873, and cultural remains from the mining era are everywhere, including a partly intact village and thousands of rock cairns and walls. Yaqui Indians from Sonora probably comprised the labor force (Bowen 2000:125-127), so technically these structures would be indigenous, but built under decidedly non-native conditions. Since guano mining stripped much of the island's surface, it probably obliterated the record of any earlier indigenous visitors.

Remarks

Rasa has been a critical nesting island for Heermann's Gulls (*Larus heermanni*), Elegant Terns (*Sterna elegans*), and Royal Terns (*Sterna maxima*) (Velarde and Anderson 1994; Velarde and others 2005). Until the island was declared a wildlife preserve in 1964, Mexican egg collectors gathered up to half a million eggs in a season for sale in towns throughout the Gulf (Bahre and Bourillón 2002:397-398; Ezcurra and others 2002:434-435). It would be surprising if the Seris did not exploit this phenomenal food supply so close to their home territory. But this makes the Seri name for the island all the more perplexing because there is no historical record of pelicans nesting there. Possibly, Rasa was a pelican nesting site prior to the disruption caused by guano mining, and after mining activity subsided the island was recolonized by gulls and terns.

Isla Partida Norte (Cardonosa)

Water Resources   None known.

Documentary Record

WJ McGee (1898:49) implied that Seris might occasionally have visited Isla Partida Norte.

Oral History

The traditional Seri name for the island is *Hast Siml,* 'Barrel Cactus Mountain' (Moser and Marlett 2005:352). Since Partida Norte has no barrel cactus, the name might be derived from a perceived resemblance between the island's small squat cardons (*Pachycereus pringlei*) (Fig. 3.29) and the familiar Sonoran barrel cactus *Ferocactus tiburonensis*. Apparently, however, the Seris have no recollection of early visits.

Archaeological Record

Extensive surveys totalling about three days in

2004 and 2005 have covered about 70 percent of the island, including most of its accessible terrain. Recorded remains include two camps, a rocky hillside with talus pits (Fig. 3.30), and a thin scatter of artifacts throughout the island. The largest camp includes two corralitos, two metates, four manos (Fig. 3.31), and several flaked stone tools and manufacturing debris. Among the isolated artifacts are a well-used metate and a bifacially-flaked *Dosinia* shell (Bowen:field notes).

Remarks

Because of the absence of ground predators, Isla Partida Norte provides crucial habitat for Least Storm-Petrels (*Halocyptena microsoma*) and Black Storm-Petrels (*Oceanodroma melania*), which nest among the island's talus blocks (Velarde and others 2005:455-456, 459). The island's talus slopes also serve as day-roosts for the Gulf's endemic fish-eating bat, *Myotis vivesi*. It could be that Partida Norte's talus pits are the result of excavations to extract these animals for food (Bowen 2000:347-348). However, the island's main attraction for native people would likely have been its dense stands of dwarf cardon cactus, which would have provided a bonanza of food and fluid in summer, when the fruit was ripe.

Figure 3.29. Isla Partida Norte, central part. One of the island's dense stands of cardon cactus. At the time of the photo, nearly all the cardons were afflicted with a bacterial infection. Although the infection was killing existing stems, most plants were generating new stems from the base. May 2005.

Figure 3.30. Isla Partida Norte, southern part. Two talus pits in a steep talus slope, bottom center and right center. The pit at the bottom center is 80 cm in diameter and 40 cm deep. Rocks removed during excavation of each pit were piled on the downhill edge. Photograph looks northeast. March 2004.

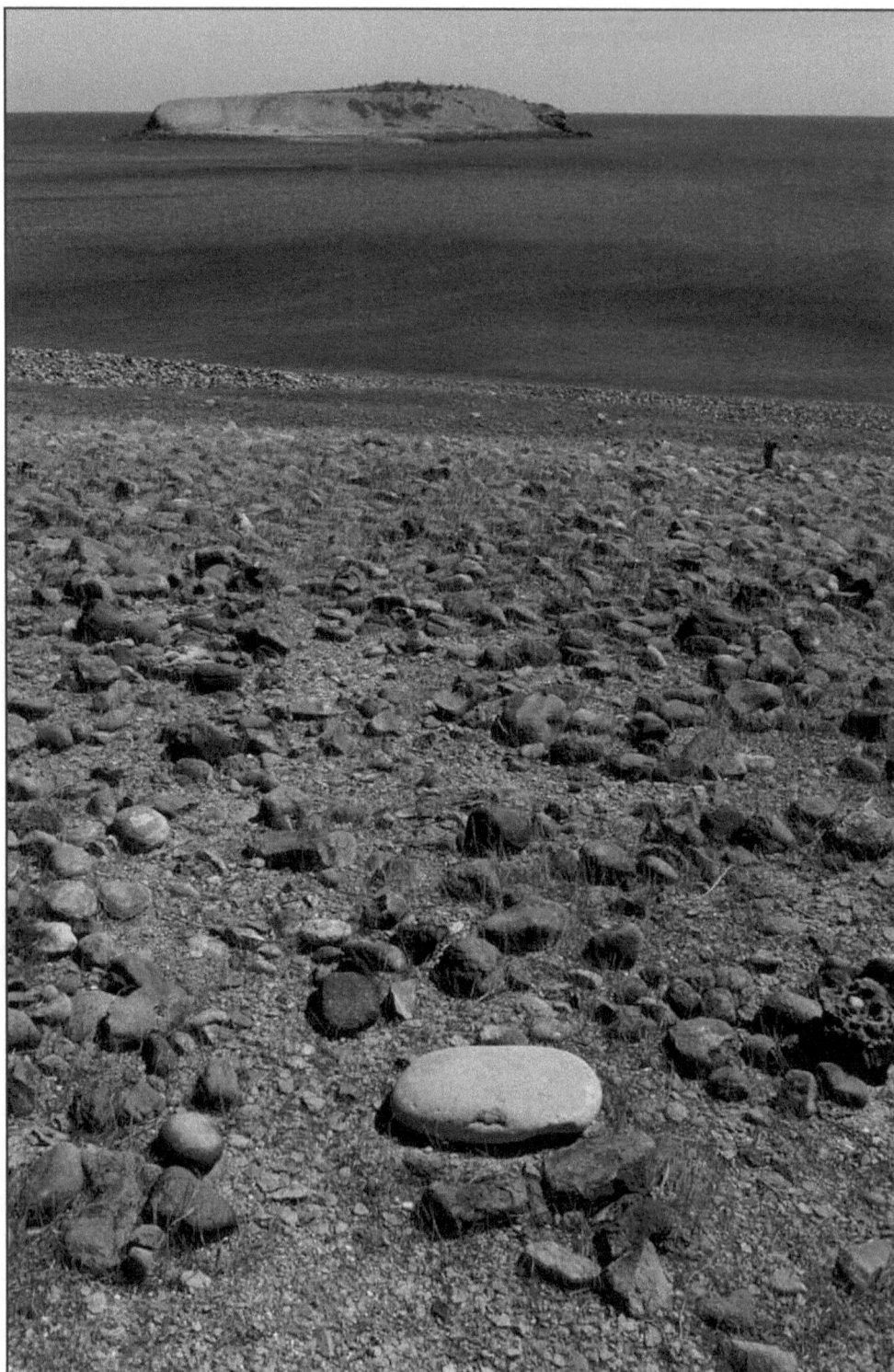

Figure 3.31. Isla Partida Norte, eastern coast. Metate and mano pair in situ, 50 cm from one another. The metate is the large light gray water-worn beach rock at lower center; the mano is the beach cobble to its left with the light gray grinding face exposed. The metate is heavily ground on both faces; the mano is unifacial and shows moderate grinding. Metate length is 35 cm, width 22 cm, thickness 7.3 cm; mano length is 13 cm, width 10 cm, thickness 5.6 cm. Photograph looks east, with Isla Cardonosa Este in the distance. February 2004.

Isla Cardonosa Este

Water Resources None known.

Documentary Record None known.

Oral History None known.

Archaeological Record

An intensive, systematic, and complete survey of this tiny island in 2004 failed to turn up a single indigenous artifact or structure (Bowen:field notes). Given its proximity to Isla Partida Norte, it is hard to imagine that native people did not even explore it. More likely, they found nothing there that warranted return visits, or their activities produced no material remains, or none that have survived.

Isla Ángel de la Guarda

Water Resources

Many commentators have expressed opinions about whether Isla Ángel de la Guarda has water, based more often on hearsay than direct observation. When Francisco de Ortega anchored there in 1636 he probably became the first European, and one of the few, to see water for himself:

> There are wells of brackish water on [this island], which is what the Indians drink. In a ravine on the eastern shore, we found water in some pools, which collects there when it rains (Ortega 1970c:465).

In 1765 Father Wenceslaus Linck explored the island for three days and:

> did not find even a water hole (Barco 1967:27).

William Strafford, writing in 1746, stated

that the island:

> has not been reconnoitered because of the lack of water (Stratford 1958:60).

Carl Beal, who visited the island in 1921, found:

> no water excepting that which lodges in the tinajas (Beal 1948:23).

According to Howard Gentry:

> No permanent source of fresh water is known (Gentry 1949:93).

As Gerhard and Gulick put it:

> As far as is known there is no fresh water on the island except that found in *tinajas*, or natural tanks in the mountains (Gerhard and Gulick 1970:211).

While it *may* be true that the island's water is limited to non-permanent tinajas, these waterholes occur throughout much of the island and constitute a major resource. Responsible reports by both scientists and local fisherman attest that some tinajas hold water for months after a rain (for example, Arnold 1957:247; Loreto Fuerte, personal communication 2005; José Smith, personal communication 2007; Charles Sylber, personal communication 2007). A flight over parts of the island in September 2006, about ten days after Tropical Storm John passed through, revealed about 60 tinajas filled to the brim with water (Fig. 3.32) and two arroyos with short segments of flowing water. A ground survey three months later established that one set of 18 of the tinajas observed from the air still contained about 16,000 liters of water, and measurements showed their maximum capacity to be about 50,000 liters (see Fig. 2.3). Other tinajas have been discovered elsewhere on the island during the course of archaeological survey (Bowen:field notes).

Documentary Record

In the spring of 1636, Francisco de Ortega went ashore at a large island he named San Sebastián, probably today's Isla Ángel de la Guarda:

> When we dropped anchor on this island, fully fifty Indians without any weapons came to the beach. From the harbor on the island's western side, we brought the boat into shore. The Indians came up to us with great fear, throwing dirt into the air, which is a sign of peace among them. [They are] a different nation from [the] others we have seen up till now (Ortega 1970c:464-465).

In 1765 Indians from the mission of San Borja reported seeing fires on Isla Ángel de la Guarda. This prompted Father Wenceslaus Linck to spend three days on the island looking for Indians and water, but he saw no trace of either. According to fellow missionary Miguel del Barco, who penned the most detailed surviving account of the expedition, Linck and his companions concluded that:

> the large fires about which the Gulf Coast Indians had informed them were imaginary or, if real, had been lighted by other [Baja] California coastal natives who crossed over to the island occasionally on rafts (Barco 1967:29; see also Clavigero 1937:346).

Oral History

The traditional Seri name for Isla Ángel de la Guarda is *Xazl Iimt* 'Where the Puma Lives' (Marlett and Moser 2005:583). In 1921, one Seri told naturalist Charles Sheldon that "coyotes, jaguars, lions and deer were plentiful" on

Figure 3.32. Isla Ángel de la Guarda, eastern side. Aerial view of a group of 20 tinajas (dark splotches) in an arroyo, photographed about ten days after Tropical Storm John filled them with water. September 2006.

the island (Sheldon 1979:116), but Seris today say the island's name makes no sense because they do not believe there are any *"tigres"* there (Stephen Marlett, personal communication 2009). During the 1960s, Edward Moser found no specific recollections of early Seri voyages to the island, although at that time Seris knew that people had occasionally gone there (Felger and Moser 1985:311). Recently, linguist Stephen Marlett elicited hints of such voyages to the northwestern tip of the island, a region where some Seris say there are sea creatures with long giraffe-like necks and strange characteristics that only a few people have ever seen (Stephen Marlett, personal communication 2004). On a more human level, the Seris say that a disjunct band of Seris, who actually lived on the Baja California peninsula, made frequent use of Isla Ángel de la Guarda. These people are known as the *Hant Ihiini Comcáac* 'Baja California Seris' or *Hast Quita Comcáac* 'Those Seris Who Had *Hast Quita* As Their Birthplace,' the name *Hast Quita* referring to a pyramidal mountain on the peninsula. It is said that these Seris made frequent trips to Isla Ángel de la Guarda to hunt rattlesnakes, which they killed in large numbers and ate with great relish (Herrera 2009).

Archaeological Record

Extensive surveys of this huge island were conducted in 1988 and 1989, and again from 2004 to 2009, totalling about 75 days of field work in nine widely-separated localities. Results consist of some 125 sites plus many scattered artifacts and structures (Bowen: field notes). Habitation sites are open-air camps, including one with about 25 clearings and corralitos, but no rockshelters. Two sparse shell middens, also presumably habitation sites, contain bones of fish, sea lions, and sea turtles, as well as shells, flaked stone artifacts, and metates and manos. Many sites of unknown function consist of up to several hundred piled rock cairns (Fig. 3.33). In some cases, cairns are spread over entire hills. Cairn sites occasionally include corralitos (Fig. 3.34) but not artifacts. Stone circles, most with closely-spaced rocks (Fig. 3.35), are widespread and in some places abundant. Small quarry-workshop sites, where stone tools were roughed out for local use, are common. Rocks quarried include rhyolite, chalcedony, and quartz. Flaked stone artifacts from these workshops are simple core and flake tools. They include agave knives, large oval and leaf-shaped bifaces, some thinned, and domed scrapers. Recognizable artifacts also include two quartz projectile points, both small leaf-shaped specimens (Fig. 3.36), a type that is apparently chronologically undiagnostic in the region (Hyland 1997:298-300). Metates are common throughout the island, but manos are scarce. No burials, rock art, or pottery have been found.

Perhaps the most spectacular sites are huge quarry-workshops associated with obsidian deposits concentrated in one small part of the island. This resource consists of many individual deposits, some more than one hectare in extent. Most were intensively worked, creating massive amounts of debitage, many bifaces broken in various stages of reduction (Fig. 3.37), and a few broken small leaf-shaped projectile points. The fact that raw obsidian, debitage, and finished obsidian tools are rare elsewhere on the island strongly suggests that expeditions to Isla Ángel de la Guarda were undertaken specifically to exploit and export this resource (Bowen:field notes).

Remarks

The longstanding belief that Isla Ángel de la Guarda is waterless and was rarely visited is clearly incorrect. It is well-watered, at least periodically, and was extensively utilized by native people.

Figure 3.33. Isla Ángel de la Guarda, southern end. Two of the nearly 250 rock cairns that lie along the ridges and summit of a prominent peak. Dimensions not recorded, but the cairn on the right is approximately 1 m in diameter and 60 cm high. Photograph looks east. January 2006.

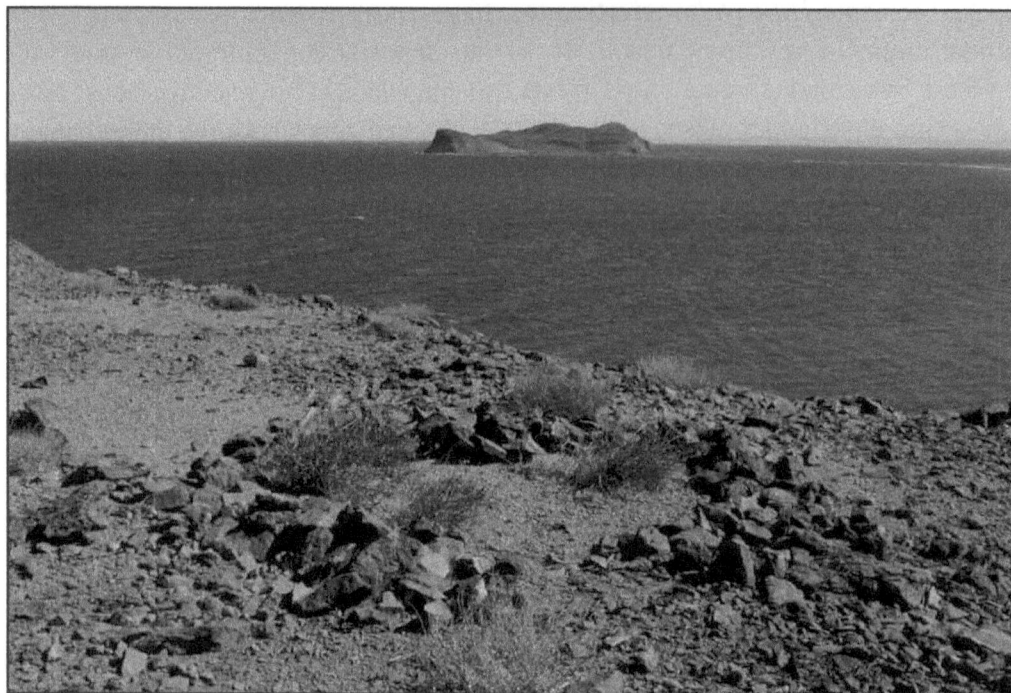

Figure 3.34. Isla Ángel de la Guarda, southern end. Corralito on a low hill with a view of Isla Estanque. It is 2.5 m long and 1.6 m wide, inside dimensions, with rocks piled up to 40 cm high. The opening, at the bottom center of the photo, is 70 cm wide and faces southwest. The site includes at least three other corralitos, four talus pits, and some 234 rock cairns. Photograph looks east-northeast. January 2006.

Figure 3.35. Isla Ángel de la Guarda, western coast. Stone circle of closely-spaced stones, half of which touch or nearly touch one another. The circle is 1.5 m long and 1.4 m wide, inside dimensions. The opening, 60 cm wide, faces east. This circle is one of at least 34 within an area about 1500 m by 500 m. Photograph looks southeast. January 2009.

Figure 3.36. Isla Ángel de la Guarda, eastern coast. Leaf-shaped projectile point of crystalline quartz from an open-air camp. Length is 5.3 cm, width 2.0 cm, thickness 8 mm. January 2007.

Figure 3.37. Isla Ángel de la Guarda, western coast. Broken obsidian biface at a quarry-workshop site associated with a typical surface deposit of obsidian nodules and pebbles. The left half lies in situ; the right half was found 1 m away. Length of the reconstructed artifact is 25.0 cm, width 7.3 cm, thickness 2.7 cm. The large number of phenocrysts is typical of most obsidian on the island. The artifact probably broke at the same time a large thinning flake struck from the reverse destroyed the upper edge. January 2009.

ISLA ESTANQUE (LA VÍBORA, POND ISLAND)

Water Resources   No data.

Documentary Record   None known.

Oral History

The traditional Seri name for Isla Estanque is *Hast Xtaasi quih iti Ihíij* 'Mountain Which Has the Estero' (Moser and Marlett 2005:352), so named for the nearly enclosed lagoon on the island's western shore.

Archaeological Record

In 1921, ornithologist Virgil Owen collected shells that were "possibly in part from a kitchen midden" (Hanna and Hertlein 1927:149).

About 20 percent of the island was surveyed during a half day of field work in 2006. At that time three corralitos, a rock cairn, a rock cluster (Fig. 3.38), a heavily-used metate (Fig. 3.39), three well-used manos, a core, about 15 flakes, and a few weathered oyster shells were recorded (Bowen:field notes).

ISLA MEJÍA

Water Resources   No data.

Documentary Record   None known.

Oral History   None known.

Archaeological Record

Extensive surveys totalling about two days between 2005 and 2007 have covered about 50 percent of the island's accessible terrain, but with few results. Indigenous remains appear to be limited to a single well-defined corralito (Fig. 3.40), two questionable corralitos (possibly natural features), two stone flakes, and two *Lyropecten* shells (Bowen:field notes).

ISLA PIOJO

Water Resources   None known.

Documentary Record   None known.

Archaeological Record

Extensive surveys totalling about two days of field work between 2004 and 2009 have covered about 50 percent of the island, including nearly all of the accessible terrain. Archaeological remains include at least 12 unusually small stone circles (Fig. 3.41), a rock cluster, a piled rock cairn, a thin scatter of debitage from the manufacture of stone tools (Fig. 3.42), five well-used manos, and two shells (Bowen:field notes).

ISLA CORONADO (SMITH)

Water Resources   No data.

Documentary Record   None known.

Oral History   None known.

Archaeological Record

The island has never been formally surveyed. An hour spent on the summit plateau in 2006 revealed three rock cairns and a corralito (Fig. 3.43) that appear to be indigenous structures (Bowen:field notes).

Figure 3.38. Isla Estanque, western side. Isolated rock cluster 1.2 m in diameter. Photograph looks south-southwest, with Isla Ángel de la Guarda in the distance. January 2006.

Figure 3.39. Isla Estanque, western side. Unifacial metate, not in situ. Photograph shows the buried face with its heavily ground basin 3 mm deep. Length is 29.0 cm, width 20.5 cm, thickness 7.4 cm. January 2006.

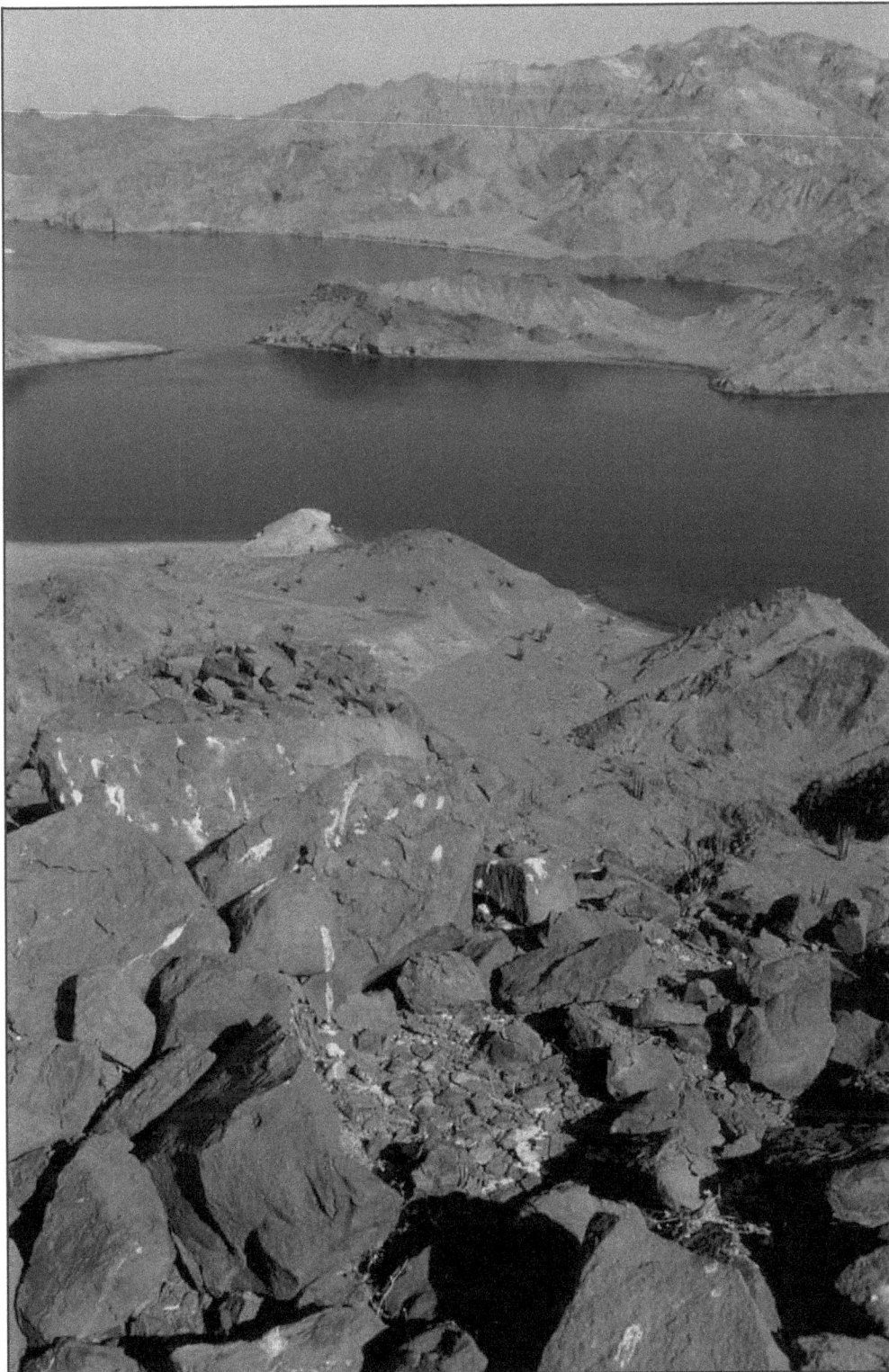

Figure 3.40. Isla Mejía, southwestern side. Small corralito on a high ridge, elevation about 230 m. The structure was positioned to incorporate a bedrock outcrop and large boulder as part of the wall, and the interior was cleared of large rocks. Dimensions not recorded. Photograph looks southeast, with Isla Ángel de la Guarda in the distance. May 2007.

Figure 3.41. Isla Piojo, southern end. One of eight small stone circles within an area about 100 m in diameter. This one consists of 15 rocks. Outside dimensions are 80 cm by 70 cm; inside dimensions 30 cm by 25 cm. The small dark object 1 m above the left side of the structure is a flaked core or chopper (see Fig. 3.42); a large limpet shell lies 1 m beyond the artifact. This circle and several others nearby have now been largely destroyed, perhaps by a treasure hunter, who scattered the rocks and excavated a small pit in the center. Photograph looks southeast. May 2004.

Figure 3.42. Isla Piojo, southern end. Bifacially-flaked core or chopper made from a basalt beach pebble, in situ. It lies 1 m from the small stone circle shown in Figure 3.41. Length is 6.4 cm, width 5.9 cm, thickness 3.9 cm. May 2004.

Figure 3.43. Isla Coronado (Smith), northern end. Corralito, foreground, on the island's summit plateau at more than 400 m elevation. The structure is 2.8 m long and 1.1 m wide, inside dimensions. Several other structures are in the immediate vicinity, although some may be modern. Photograph looks west-southwest. May 2006.

Isla San Luis (Encantada Grande, Salvatierra)

Water Resources	None known.
Documentary Record	None known.
Oral History	None known.

Archaeological Record

Extensive surveys totalling about one full day of field work between 2004 and 2006 have covered approximately 25 percent of the island, including about half of its accessible terrain. Recorded remains consist of at least one flaked chopper (Fig. 3.44), two stone circles (Fig. 3.45), and four other flaked stone artifacts (Bowen:field notes).

Remarks

Nomenclature for this island has been confused. On maps it is labelled "San Luis," the name adopted here. In the 1940s local fishermen called it "La Encantada Grande," and today area residents unanimously refer to it as "Isla Salvatierra." (see discussions by Klauber [1949:96] and Cliff [1954:70]).

Figure 3.44. Isla San Luis, central part. Bifacially-flaked chopper of local tabular volcanic rock, located 200 m from the stone circle shown in Figure 3.45. Length is 19.2 cm, width 16.5 cm, thickness 3.2 cm. Both ends have been dulled from use. May 2005.

Figure 3.45. Isla San Luis, central part. Stone circle of widely-spaced stones, 1.1 m long and 1.0 m wide, inside dimensions. Photograph looks west, with Baja California in the distance. May 2005.

Chapter 4
Summary and Implications

SUMMARY

The previous chapter considered current evidence, or lack thereof, of native people on 32 Gulf islands. This list includes all 22 islands larger than 2 km^2 and ten smaller islands for which some data exist (Table 3.2). Collectively, the data suggest that native people were familiar with at least 29 of the 32 islands. For 19 of the islands, the data can be considered unequivocal, consisting of unambiguous historical documentation, credible oral history, and/or a strong archaeological record. For ten islands the evidence is limited, weak, or equivocal, and hence in need of corroboration. There are no data of any kind for two islands. One small island, which has been subjected to an intensive and complete archaeological survey, has produced no evidence of native people (Table 4.1).

Not surprisingly, Indians exploited some islands more than others. There is an obvious, if loose, correlation between island size and evidence of occupation. This is presumably a function of the well-known relationship between island area, physiographic (and hence niche) diversity, and species richness. Large islands tend to have more resources of all kinds and, importantly, are more likely to have fresh water. In fact, nine of the ten largest have permanent or long-term water resources. Thus it is no surprise that most of the larger islands supported communities of Indians—permanently, seasonally, or intermittently. This was certainly the case for Cerralvo, Espíritu Santo/Partida Sur, San José, Tiburón, and San Esteban, and probably true for Santa Catalina, Carmen, San Marcos, and Ángel de la Guarda. The only island of the ten largest that may not have sustained a population on a frequent basis was San Lorenzo, but even there Seri oral history and archaeology suggest sporadic visits over an extended span of time.

Small islands without water or a broad diversity of plants and animals may have had specific resources that drew native visitors seasonally or opportunistically. Some of the smallest islands, including San Ildefonso, San Pedro Mártir, Alcatraz, Patos, Rasa, San Luis, and perhaps Partida Norte support, or once supported, huge colonies of nesting seabirds that would have constituted a virtually unlimited food supply every spring (Anderson and Keith 1980:75; Velarde and others 2005:Table 23.1; Velarde and Anderson 1994:Table 1). Similarly, the cardon forests of Islas San Pedro Mártir and Partida Norte would have offered an abundance of tasty fruit each summer. Sea lions, easily hunted, maintain rookeries or non-breeding colonies on several islands, including Ángel de la Guarda, San Esteban, and the small islands of San Pedro Nolasco, San Pedro Mártir, and Patos (Le Bouef and others 1983:Table 1; Underwood and others 2008), and some of

Table 4.1. Evaluation of evidence of native people based on all sources of data

Evidence Unequivocal	Evidence Needs Corroboration	No Evidence	No Data
Cerralvo	Monserrat	Cardonosa Este	Santa Cruz
Espíritu Santo/	Danzante		Tortuga
Partida Sur	Los Coronados		
San Francisco	San Ildefonso		
San José	San Pedro Nolasco		
Santa Catalina	San Pedro Mártir		
Carmen	Patos		
San Marcos	Rasa		
Alcatraz	Mejía		
Tiburón	Coronado (Smith)		
Dátil			
San Esteban			
San Lorenzo			
Las Ánimas			
Salsipuedes			
Partida Norte			
Ángel de la Guarda			
Estanque			
Piojo			
San Luis			

these colonies were apparently enormous in the historic past (Bowen 2000:102-105). Occasional storms in the upper Gulf transform Islas Las Ánimas and San Luis into veritable carpets of amaranth (*Amaranthus watsonii*) whose seeds were a major resource for Indians on both sides of the Gulf (Aschmann 1959:86-87; Felger and Moser 1985:228-229). And of course, it is important to bear in mind that, for some islands, the resources we see today may have been greatly reduced by the excesses of Mexican/European/North American exploitation during the past century and a half.

Distance seems to have been no barrier to native use of the islands. In the hands of Gulf Indians, the balsa was a swift and seaworthy craft. Ortega (1970b:449) noted in 1633 that the Indians around La Paz traveled "four to six leagues [10 to 15 km] out from shore to fish." These were presumably day trips and would have entailed round trip distances of 20 to 30 km. In the late 1640s Pedro Porter y Casanate (1970:892) wrote that the Indians of the southern Gulf propelled their balsas "with great speed." A century later Manuel Correa (1946:554) wrote of Seri balsas that they "slice through the water very rapidly." WJ McGee, whose party actually used a Seri balsa in 1895, was astounded by its superb performance (see p.19). In addition to efficient watercraft, the Seris (and no doubt Indians throughout the Gulf) could draw on expert knowledge of winds and currents to further speed their trips and insure their safety (Hills 2004). Even assuming an unrealistically slow average speed of 3 km per hour, a 30-km voyage would have required just ten hours. Thus no island in the Gulf would have been more than a day's paddle from the peninsula or mainland except Tortuga and San Pedro Mártir, and the overwater distance to all other remote islands could have been reduced to less than 21 km by island hopping.

In Baja California, balsas largely dropped out of use during the eighteenth century, and thus it is easy to underestimate their importance to peninsular Indians. The earliest documentary sources offer clear testimony to their widespread and regular service in both subsistence and transportation, and this record persists into the writings of the Jesuits. However, as the Indians were enticed away from the islands, concentrated in the mission settlements, and eventually decimated by European diseases or assimilated into the mestizo population, balsas largely disappeared from the documentary record. Thus once again, the best information comes from the Seris. In two rare quantitative statements, Robert Hardy in 1826 saw "fifteen or twenty" balsas off the eastern coast of Isla Tiburón (Hardy 1977:291), while Tomás Spence counted 97 balsas during a circumnavigation of the island in 1844 (Spence 1985:152). The present Seris maintain that, for the people of Isla San Esteban, the balsa was an item of daily use and every man owned one. The San Esteban people entrusted their very lives to the balsa, because whenever the water supply on San Esteban failed the entire population had no choice but to paddle to Isla Tiburón (Bowen 2000:21-22). When food ran low on San Esteban, it is said that these same people would often paddle in the opposite direction to Isla San Lorenzo, a fact recently corroborated by archaeological evidence (Bowen 2005a). For the Seris generally, the entire constellation of Midriff Islands was readily accessible by island hopping, and their traditional names and knowledge of these islands suggest that they knew them all when balsas were still in use.

There are also hints of much longer voyages. According to Seri oral history, the San Esteban people sometimes paddled to Baja California, and on one occasion staged a raid on Mulegé, some 200 km south of Isla San Esteban (Bowen 2000:24-25). Conversely, it is said that the Seri group known as the *Hant Ihiini* people,

who lived on the Baja California peninsula, made frequent foraging trips to Isla Ángel de la Guarda and occasionally embarked on trading expeditions to Isla Tiburón (Herrera 2009). The historic literature also contains suggestions of extended Seri voyages by balsa. Seris were suspected in an 1802 raid on Santa Gertrudis, a mission settlement on the peninsular side of the Midriff region, and they were implicated in a series of early nineteenth century raids on the town of Loreto, more than 300 km to the south (Bowen 2000:232-234).

Although the Isleños of the southern Gulf traveled frequently between the peninsula and their three major islands, it is unclear whether peninsular people engaged in long distance travel to the same extent as the Seris. The longest voyage currently known was intended to take people from Isla San José to Loreto, a distance by balsa of about 130 km, although the Indians were forced to turn back at Isla Santa Catalina because of unfavorable winds (Bravo 1970:26; see p. 34).

If these sources are even remotely reliable, balsa traffic throughout the Gulf must have been extensive when Europeans first arrived. How far back the balsa extends into prehistory is anybody's guess, as it is unlikely that any will ever be found. Whether ultimately cast adrift, left to rot on beaches, broken up by storms, or burned in funerary rites (Griffen 1959:28), balsas are among the least likely artifacts to survive as part of the archaeological record.

CHRONOLOGY

The question of the antiquity of the balsa raises the more general question of island chronology. Over what span of time did native people exploit the islands? When did people first travel to the islands?

At present, island chronology is sketchy.

Obviously, historical documents all post-date the arrival of Europeans, and the historic record of the islands effectively begins with Ulloa's voyage of 1539. For the native people of the southern islands, documentary sources provide specific information mainly for the one-hundred-year interval between Ortega's 1632 voyage and the 1730s, when most of those islands were abandoned. For the Midriff Islands, the written record of native people begins with Cardona's 1615 encounter with the Seris on Isla Tiburón and Ortega's 1636 encounter with Indians on Isla Ángel de la Guarda. Although subsequent documentation of Seris on Isla Tiburón has been more or less continuous since 1615, there is precious little historic record of Indians on other Midriff islands, and the little information that does exist comes mostly from the nineteenth century.

Seri oral history firmly places Seris on several Midriff islands besides Tiburón, but the time depth for recalling specific events probably does not extend beyond the early nineteenth century. That the Seri have traditional names for all the major Midriff islands, most of them descriptive of physical features, suggests that their familiarity with them may reach much deeper into the past, but there is no way at present that this can be demonstrated.

Archaeology is the one approach capable of extending island chronology into the prehistoric past, but there has been only limited progress on this front. Fewer than half of the 32 islands in the sample considered here have been systematically surveyed, and controlled excavations have been conducted on only three (Cerralvo, Espíritu Santo/Partida Sur, and San Esteban). The archaeological record for many islands consists only of surface remains, with few or no time-sensitive artifacts or structures, and no charcoal, bone, wood, or other organic material suitable for radiocarbon analysis.

Radiocarbon dates have been secured for only five islands. These are Cerralvo, Espíritu Santo/Partida Sur, San José, Tiburón, and San Esteban. Isla Cerralvo has yielded two dates, and Isla San José one. All three fall in the late prehistoric period between AD 700 and 1270. The four radiocarbon dates from Isla San Esteban all post-date AD 1650, which is consistent with the island's few time-sensitive artifacts and Seri oral history. The single uncalibrated radiocarbon age from Isla Tiburón of 1100 ± 300 BP is consistent with the island's intrusive pottery, but the presence of presumed Paleoindian points suggests that Isla Tiburón may have been inhabited since at least the early Holocene.

At present, Isla Espíritu Santo/Partida Sur is the only island with a suite of radiocarbon dates and a demonstrably long occupational sequence. Samples obtained from several sites span the Holocene, from about 9000 BC to the fifteenth century AD. Radiocarbon ages on shells from the lowest levels of Covacha Babisuri range from approximately 36,550 to greater than 47,500 BP, but it seems likely that these shells were already ancient when people collected them (see p. 28).

During the Pleistocene-Holocene transition, the Gulf was a very different place, and not all of today's islands existed, or existed as islands. It is possible that some young volcanic islands, particularly San Ildefonso, Tortuga, Rasa, and San Luis, had not yet fully emerged from the sea (Carreño and Helenes 2002:Table 2.1). Others existed as land masses but not as islands. During the last glacial maximum, about 21,000 to 19,000 years ago, so much sea water was tied up on land in continental ice sheets and glaciers that global sea level was depressed by at least 120 m to 135 m below its present stand (Clark and Mix 2002:5-6; Peltier 2002:377-378, 391-392). Some of today's Gulf islands were connected to the peninsula or mainland by dry land, and if any people were present at that time, they could have reached them simply by walking. They never would have suspected that these chunks of land would become islands, and that their distant descendants would need watercraft to reach them.

Beginning about 19,000 BP the ice sheets began to retreat. Melting was rapid and sustained between about 16,000 and 12,500 BP, then stalled for a thousand years during the Younger Dryas cold interval, and finally resumed about 11,500 BP. By about 7000 BP, the northern ice sheets had retreated to their present state, and further melting over the next thousand years was confined largely to the Antarctic ice sheet (Lambeck and Chappell 2001:682-683; Peltier 2002). As the ice melted, global sea level rose, reaching about 60 m below its present position by 15,000 BP, 50 m by about 12,500 BP, 40 m around 10,000 BP, and attaining its modern stand by about 6000 BP (Davis 2006:18-19).

The insular status of present-day Gulf islands as the sea rose was affected by numerous factors besides changing sea level. Many of these are plagued with uncertainties, not the least of which is grossly inadequate near-shore bathymetry (Davis 2006:23; Murphy and others 2002:450; also compare Dauphin and Ness 1991:Plates 1 and 2 with Carreño and Helenes 2002:Table 2.2, Lindsay and Engstrand 2002:Tables 1.1-1.3, and Murphy and others 2002:Table 1.1-1). That said, it would appear that even during the sea level lowstand of the last glacial maximum, half or more of the 32 islands considered here were already isolated by water and could only have been reached with watercraft. In the southern Gulf, these include Islas Cerralvo, Santa Cruz, Santa Catalina, Tortuga, and probably Monserrat. In the Midriff, they include San Pedro Nolasco, San Pedro Mártir, San Esteban, and the entire archipelago from San Lorenzo to Ángel de la Guarda. By 12,500 BP, rising sea level may have isolated some islands that formerly had been connected to shore by land bridges, and between 12,500 and 6000 BP landbridge con-

nections to all the remaining islands were presumably severed. Thus while the first people in the Gulf region might have reached some landbridge islands on foot, after about 6000 BP none of the present Gulf islands was accessible without watercraft. Turning the statement around, archaeological remains that post-date 6000 BP on any present-day island can be taken as *de facto* evidence of watercraft. And for non-landbridge islands, *any* archaeological remains constitute evidence of watercraft.

Clovis fluted projectile points from both sides of the Gulf suggest that humans were established on both the peninsula and mainland by about 13,000 calendar years ago (Hyland and Gutiérrez 1996, Sánchez 2001; Waters and Stafford 2007) and there is a record of more or less continuous occupation after that time (Carpenter and others 2005; Laylander and Moore 2006). So far, the only evidence of terminal Pleistocene-early Holocene occupation of the Gulf islands comes from Isla Espíritu Santo/Partida Sur, with its suite of radiocarbon dates, and less certainly from Isla Tiburón, with its apparent Paleoindian projectile points. On the Pacific side of the peninsula, Isla Cedros (Fig. 1.1) has recently produced a series of radiocarbon dates ranging from about 9200 to 12,000 calendar years BP plus the base of a Clovis point (Des Lauriers 2006; 2007). It may not be coincidental that these three islands—Cedros, Tiburón, and Espíritu Santo/Partida Sur—all had landbridge connections at the end of the Pleistocene that would have made them accessible by foot. It will be interesting to see if future archaeological investigations turn up equally early material on islands that could only have been reached with watercraft.

BROADER HORIZONS

One cannot reasonably consider the human history of the Gulf islands without broaching the subject of the peopling of the Americas. Of particular relevance is the hypothesis of human entry into North America along the Pacific rim by coastally-adapted people equipped with sturdy watercraft (Anderson and Gillam 2000; Fedje and others 2004; Fladmark 1979; 1983; Goebel and others 2008; Gruhn 1988; 1994). In bringing up this hypothesis, my intent is not to argue its virtues or shortcomings, but simply to point out the potential significance of the Gulf of California, based on a scenario recently developed by R. James Hills and outlined here with his permission (Hills, personal communication 2008; see also Hills and Yetman 2007:512-513).

Many versions of the coastal entry hypothesis are concerned only with getting late Pleistocene voyagers past the continental ice sheets to what is now the Canada-United States boundary (Clague and others 2004; Fladmark 1979; 1983; Gruhn 1994; Mandryk and others 2001). In versions that take people farther south, it is usually assumed that once the voyagers had outflanked the ice they (or some of them) dispersed along the Pacific coast and on down the Baja California peninsula to Cabo San Lucas. From there, it is further assumed (albeit vaguely) that they crossed the mouth of the Gulf to the Mexican mainland, making landfall somewhere between southern Sinaloa and Jalisco, before continuing southward (for example, Anderson and Gillam 2000:53; Brace and others 2004:35; Dixon 1999:33; 2001:292; Pearson 2004:95).

Thus, in these extended coastal scenarios, early voyagers would have bypassed the Gulf of California altogether. However, Hills points out that voyages across the mouth of the Gulf are highly unlikely. From the southern tip of the peninsula, the Mexican mainland is not only some 280 km away, but the broad low-lying coastal plains are totally invisible, even from a high vantage point. People arriving at the southern tip would have seen empty ocean

in all directions except along the peninsula itself. Hills maintains that no coastally-adapted people in their right minds would have struck out into this apparently empty and endless sea. The first people to arrive at Cabo San Lucas were not "headed" anywhere, and their only objective would have been to keep to the productive coastal waters that provided them with their livelihood. Consequently, Hills argues that they almost certainly would have rounded the cape and dispersed northward into the Gulf, following the eastern coast of Baja California, and reconnoitering the islands along the way.

The first hint that these people would have had of land to the east would probably have come some 500 km up the Gulf in the vicinity of present-day Santa Rosalía, where the mountains near Guaymas are visible on a clear day. However, the Gulf at that point is about 130 km wide and, as Hills notes, people subsisting well on the resources of the peninsular coast would have no reason to cross such an expanse of open water. But the Midriff region, another 140 km up the coast, is a different story. Although the Gulf there is about 92 km wide, one can island hop from the peninsula to the mainland via Islas San Lorenzo, San Esteban, and Tiburón, with no overwater crossing exceeding 17 km. Moreover, until sea level neared its modern stand, the distance from the peninsula to the mainland was only about 60 km because Isla Tiburón *was* the mainland. Historical accounts of balsa travel suggest that crossing the Midriff by way of the islands would have required no more than three days of paddling, with land always in plain sight.

The commercial sea kayak is the modern analog of the indigenous balsa. Today, skilled kayakers bent on a good adventure make the same island-hopping voyage between the peninsula and the mainland. In the historical past, small groups of Seri men from Isla San Esteban sometimes set off in their balsas on similar ventures, with no objective other than exploration—and maybe a little raiding (Bowen 2000: 24-25). One can readily envision much the same scenario playing out soon after the first people arrived in the Midriff—small groups of adventurous young men setting out to explore the islands to the east and arriving on the mainland a few days later. Once people reached the mainland, the way south was a relatively straight shot all the way to Tierra del Fuego. In Hills' scenario, the Gulf would not be a remote backwater in the peopling of the Americas, but a pivotal region in the dispersal process.

Whatever the merits of coastal entry and dispersal hypotheses, the islands are unquestionably important for reconstructing the prehistory of the Gulf of California. Lowered sea level during the late Pleistocene-early Holocene shifted continental shorelines to the seaward side of landbridge islands, making these islands obvious places to search for sites of early coastal people. Sites of early people on non-landbridge islands would be equally important because they would provide implicit evidence of watercraft. And from a practical standpoint, islands may be good places to look because their relative isolation in modern times has spared some of them from wholesale site destruction by development and looting. The value of locating and excavating stratified rockshelters is underscored by Harumi Fujita's work at Covacha Babisuri. She and her colleagues have done an exemplary job of elucidating the long record of native people on Islas Espíritu Santo and Partida Sur, but their work is only a beginning. Loren Davis sums it up succinctly:

> A complete archaeological survey of Baja California's islands is needed (Davis 2006:23).

One could say the same for Sonora's islands as well.

References Cited

Anderson, Daniel W. and James O. Keith
  1980 The Human Influence on Seabird Nesting Success: Conservation Implications. *Biological Conservation* 18(1):65-80.

Anderson, David G. and J. Christopher Gillam
  2000 Paleoindian Colonization of the Americas: Implications from an Examination of Physiography, Demography, and Artifact Distribution. *American Antiquity* 65(1):43-66.

Arnaud, Paul H., Jr.
  1970 *The Sefton Foundation* Orca *Expedition to the Gulf of California, March-April, 1953. General Account.* Occasional Papers of the California Academy of Sciences No. 86.

Arnold, Brigham A.
  1957 *Late Pleistocene and Recent Changes in Land Forms, Climate, and Archaeology in Central Baja California.* University of California Publications in Geography 10(4):201-317, Berkeley.

de Arrillaga, José Joaquín
  1999 [1802] [Two letters to the Viceroy of New Spain]. In *Empire of Sand: The Seri Indians and the Struggle for Spanish Sonora, 1645-1803*, edited by Thomas E. Sheridan, pp. 452-454. University of Arizona Press, Tucson.

Aschmann, Homer
  1959 *The Central Desert of Baja California: Demography and Ecology.* Ibero-Americana 42. University of California Press, Berkeley.

  1965 Historical Sources for a Contact Ethnography of Baja California. *California Historical Society Quarterly* 44(2):99-121.

Baegert, Johann Jakob
  1952 [1772] *Observations in Lower California.* Translated, with an introduction and notes, by M.M. Brandenburg and Carl L. Baumann. University of California Press, Berkeley.

Bahre, Conrad J. and Luis Bourillón
  2002 Human Impact in the Midriff Islands. In *A New Island Biogeography of the Sea of Cortés*, edited by Ted J. Case, Martin L. Cody, and Exequiel Ezcurra, pp. 383-406. Oxford University Press, New York.

Bancroft, Griffing
  1932 *The Flight of the Least Petrel.* G.P. Putnam's Sons, New York.

Banks, Richard C.
1962 *A History of Explorations for Vertebrates on Cerralvo Island, Baja California*. Proceedings of the California Academy of Sciences, Fourth Series, 30(6):117-125.

1963 *Birds of the Belvedere Expedition to the Gulf of California*. Transactions of the San Diego Society of Natural History 13(3):49-60.

del Barco, Miguel
1967 [ca. 1770s] Barco's Report on Linck's 1765 Expedition to Angel de la Guarda Island. In *Wenceslaus Linck's Reports and Letters, 1762-1778*, translated, edited, and annotated by Ernest J. Burrus, S.J., pp. 25-29. Baja California Travel Series 9. Dawson's Book Shop, Los Angeles.

1980 [ca. 1770-1780] *The Natural History of Baja California*. Translated by Froylán Tiscareno. Baja California Travel Series 43. Dawson's Book Shop, Los Angeles.

1981 [ca. 1770-1780] *Ethnology and Linguistics of Baja California*. Translated by Froylán Tiscareno. Baja California Travel Series 44. Dawson's Book Shop, Los Angeles.

Barnes, Thomas C., Thomas H. Naylor, and Charles W. Polzer
1981 *Northern New Spain: A Research Guide*. University of Arizona Press, Tucson.

Beal, Carl H.
1948 *Reconnaissance of the Geology and Oil Possibilities of Baja California, Mexico*. Geological Society of America Memoir 31, New York.

Belden, Samuel (compiler)
1880 *The West Coast of Mexico, from the Boundary Line between the United States and Mexico to Cape Corrientes, including the Gulf of California*. U.S. Hydrographic Office, Bureau of Navigation No. 56. U.S. Government Printing Office, Washington.

Bolton, Herbert Eugene
1936 *Rim of Christendom*. Macmillan, New York.

Bowen, Thomas
1976 *Seri Prehistory: The Archaeology of the Central Coast of Sonora, Mexico*. Anthropological Papers of the University of Arizona 27, Tucson.

1983 Seri. In *Southwest*, edited by Alfonso Ortiz, pp. 230-249. Handbook of North American Indians, Vol. 10, William C. Sturtevant, general editor, Smithsonian Institution, Washington.

1986 *Survey of Bahía Vaporeta, West Coast of Tiburón Island, Mexico*. Typescript on file, Centro Instituto Nacional de Antropología e Historia Sonora, Hermosillo. (Copy in Bowen's possession).

2000 *Unknown Island: Seri Indians, Europeans, and San Esteban Island in the Gulf of California*. University of New Mexico Press, Albuquerque.

2003 Hunting the Elusive Organ Pipe Cactus on San Esteban Island in the Gulf of California. *Desert Plants* 19(1):15-28, plus errata sheet in Vol. 19(2).

2004 Archaeology, Biology and Conservation on Islands in the Gulf of California. *Environmental Conservation* 31(3):199-206.

Bowen, Thomas (continued)
2005a A Historic Seri Site on Isla San Lorenzo. *Kiva* 70(4):399-412.

2005b *Archaeological Resources of Isla Alcatraz.* Typescripts on file, Guaymas, Sonora office of Área de Protección Flora y Fauna Islas del Golfo de California and Prescott College Center for Cultural and Ecological Studies, Bahía Kino, Sonora. (Copy in Bowen's possession).

Bowen, Thomas and Edward Moser
1968 Seri Pottery. *Kiva* 33(3):89-132.

Brace, C. Loring, A. Russell Nelson, and Pan Qifeng
2004 Peopling of the New World: A Comparative Craniofacial View. In *The Settlement of the American Continents: A Multidisciplinary Approach to Human Biogeography*, edited by C. Michael Barton, Geoffrey A. Clark, David R. Yesner, and Georges A. Pearson, pp. 28-38. University of Arizona Press, Tucson.

Bravo, Jaime
1970 [1720] Razón de la Entrada al Puerto de La Paz. In *Testimonios Sudcalifornios*, edited and with an introduction and notes by Miguel León-Portilla, pp. 25-67. Instituto de Investigaciones Históricas 9, Universidad Nacional Autónoma de México, México.

1979 [1717] Presenta el P. Jaime Bravo al Señor Virrey Dos Escritos: Quien los Remitte al Parescer de la Junta. In *Empressas Apostolicas de los PP. Missioneros de la Compañia de Jesus, de la Provincia de Nueva-España*, by Miguel Venegas, Libro 4, Capítulo 14. Obras Californianas del Padre Miguel Venegas, S.J., Vol. 4, edited by W. Michael Mathes, pp. 208-212. Universidad Autónoma de Baja California Sur, La Paz.

Burrus, Ernest J.
1972 Two Fictitious Accounts of Ortega's "Third Voyage" to California. *Hispanic American Historical Review* 52(2):272-283.

Cano Ávila, Gastón
ca. 1960 *The Seri Indians of the Sonora Coast.* Typescript on file, Arizona State Museum Library, University of Arizona, Tucson. (Copy in Bowen's possession).

Carabias Lillo, Julia, Javier de la Maza Elvira, David Gutiérrez Carbonell, Mario Gómez Cruz, Gabriela Anaya Reina, Alfredo Zavala González, Ana Luisa Figueroa, and Benito Bermúdez Almada
2000 *Programa de Manejo Área de Protección de Flora y Fauna Islas del Golfo de California, México.* Comisión Nacional de Áreas Naturales Protegidas, Secretaría de Medio Ambiente, Recursos Naturales y Pesca, México.

Carbonel de Valenzuela, Estevan
1970 [1632] Relación que Dió el Piloto Estevan Carbonel al Virrey de Nueva España Marques de Serralvo en México a 30 de Septiembre de 1632. In *Californiana II: Documentos para la Historia de la Explotación Comercial de California 1611-1679*, Vol. I, edited by W. Michael Mathes, pp. 338-356. Colección Chimalistac 29. Ediciones José Porrua Turanzas, Madrid.

de Cardona, Nicolás
1974 [1632] Report of the Exploration of the Kingdom of California by Captain and Commander Nicolás de Cardona. In *Geographic and Hydrographic Descriptions of Many Northern and Southern Lands and Seas in the Indies, Specifically of the Discovery of the Kingdom of California*, translated and edited by W. Michael Mathes, pp. 95-106. Baja California Travel Series 35. Dawson's Book Shop, Los Angeles.

Carpenter, John P., Guadalupe Sánchez, and María Elisa Villalpando C.
2005 The Late Archaic/Early Agricultural Period in Sonora, Mexico. In *The Late Archaic across the Borderlands*, edited by Bradley J. Vierra, pp. 13-40. University of Texas Press, Austin.

Carreño, Ana Luisa and Javier Helenes
2002 Geology and Ages of the Islands. In *A New Island Biogeography of the Sea of Cortés*, edited by Ted J. Case, Martin L. Cody, and Exequiel Ezcurra, pp. 14-40. Oxford University Press, New York.

Case, Ted J. and Martin L. Cody
1983 Preface. In *Island Biogeography in the Sea of Cortéz*, edited by Ted J. Case and Martin L. Cody, pp. vii-x. University of California Press, Berkeley.

1987 Testing Theories of Island Biogeography. *American Scientist* 75(4):402-411.

Case, Ted J., Martin L. Cody, and Exequiel Ezcurra (editors)
2002 *A New Island Biogeography of the Sea of Cortés*. Oxford University Press, New York.

Cavallero Carranco, Juan
1966 [1668] Summary Report of the Voyage Made to the Californias by Captain Francisco de Lucenilla. In *The Pearl Hunters in the Gulf of California 1668*, transcribed, translated, and annotated by W. Michael Mathes, pp. 23-85. Baja California Travel Series 4. Dawson's Book Shop, Los Angeles.

Cavazos, Tereza
2008 Clima. In *Bahía de los Ángeles: Recursos Naturales y Comunidad*, edited by Gustavo D. Danemann and Exequiel Ezcurra, pp. 67-90. Secretaría de Medio Ambiente y Recursos Naturales, México.

Chambers, George W.
1975 How Long Is a Piece of String? *Journal of Arizona History* 16(2):195-196.

Clague, John J., Thomas A. Ager, and Rolf W. Matthewes
2004 Environments of Northwestern North America before the Last Glacial Maximum. In *Entering America: Northeast Asia and Beringia before the Last Glacial Maximum*, edited by David B. Madsen, pp. 63-94. University of Utah Press, Salt Lake City.

Clark, Peter U. and Alan C. Mix
2002 Ice Sheets and Sea Levels of the Last Glacial Maximum. *Quaternary Science Reviews* 21:1-7.

Clavigero, Francisco Javier
1937 [1789] *The History of [Lower] California*. Translated by Sara E. Lake and edited by A.A. Gray. Stanford University Press, Stanford University, California.

Cliff, Frank S.
1954 *Snakes of the Islands in the Gulf of California, Mexico*. Transactions of the San Diego Society of Natural History 12(5):67-98.

Cody, Martin, Reid Moran, Jon Rebman, and Henry Thompson
2002 Plants. In *A New Island Biogeography of the Sea of Cortés*, edited by Ted J. Case, Martin L. Cody, and Exequiel Ezcurra, pp. 63-111. Oxford University Press, New York.

Correa, Manuel
 1946 [1750] Descripción de la Isla del Tiburón. In "Diario de lo Acaecido y Practicado en la Entrada que se Hizo a la Isla del Tiburón éste Año 1750," by Francisco Antonio Pimentel, pp. 552-558. *Boletín del Archivo de la Nación* 17(4):503-574. Secretaría de Gobernación, México.

Dauphin, J. Paul and Gordon E. Ness
 1991 *Bathymetry of the Gulf and Peninsular Province of the Californias.* American Association of Petroleum Geologists Memoir 47. Tulsa, Oklahoma.

Davis, Edward H.
 1922 *Trip to Tiburon Island.* Typescript of Davis' Journal, pp. 1-35. Copy in Bowen's possession.

 1934 [Trip to Alcatraz Island]. Typescript of Davis' journal, pp. 139-159. Copy in Bowen's possession.

 1965 The Pelican Skin Robe. In *Edward H. Davis and the Indians of the Southwest United States and Northwest Mexico*, arranged and edited by Charles Russell Quinn and Elena Quinn, pp. 213-218. Elena Quinn, Downey, California.

Davis, Loren G.
 2006 Baja California's Paleoenvironmental Context. In *The Prehistory of Baja California*, edited by Don Laylander and Jerry D. Moore, pp. 14-23. University Press of Florida, Gainesville.

Des Lauriers, Matthew R.
 2006 Terminal Pleistocene and Early Holocene Occupations of Isla de Cedros, Baja California, Mexico. *Journal of Island and Coastal Archaeology* 1(2):255-270.

 2007 *Remembering the Forgotten Peninsula: Theoretical and Conceptual Contributions of Archaeological Research in Baja California.* Paper presented at the 2007 meeting of the Society for California Archaeology, San Jose.

Diguet, Léon
 1898 *Rapport sur une Mission Scientifique dans la Basse-Californie.* Imprimerie Nationale, Paris.

 1973 [1905] Ancient Native Burials of Southern Baja California. Translated by Jutta Banks. *Pacific Coast Archaeological Society Quarterly* 9(1):27-30.

Dixon, E. James
 1999 *Bones, Boats, and Bison.* University of New Mexico Press, Albuquerque.

 2001 Human Colonization of the Americas: Timing, Technology, and Process. *Quaternary Science Reviews* 20:277-299.

Dixon, Keith A.
 1990 *La Cueva de la Pala Chica: A Burial Cave in the Guaymas Region of Coastal Sonora, Mexico.* Vanderbilt University Publications in Anthropology No. 38. Nashville, Tennessee.

Dockstader, Frederick J.
 1961 A Figurine Cache from Kino Bay, Sonora. In *Essays in Pre-Columbian Art and Archaeology*, edited by Samuel K. Lothrop, pp. 182-191. Harvard University Press, Cambridge.

Emerson, William K.

1960 Results of the Puritan-American Museum of Natural History Expedition to Western Mexico 12: Shell Middens of San José Island. *American Museum Novitates* No. 2013, New York.

Ezcurra, Exequiel, Luis Bourillón, Antonio Cantú, María Elena Martínez, and Alejandro Robles

2002 Ecological Conservation. In *A New Island Biogeography of the Sea of Cortés*, edited by Ted J. Case, Martin L. Cody, and Exequiel Ezcurra, pp. 417-444. Oxford University Press, New York.

Fay, George E.

1961 A Shell Circle at Puerto Kino, Sonora. *Man* 61(Article 53):56.

Fedje, Daryl W., E. James Dixon, Quentin Mackie, and Timothy H. Heaton

2004 Late Wisconsin Environments and Archaeological Visibility on the Northern Northwest Coast. In *Entering America: Northeast Asia and Beringia before the Last Glacial Maximum*, edited by David B. Madsen, pp. 97-138. University of Utah Press, Salt Lake City.

Felger, Richard S. and Charles H. Lowe

1976 *The Island and Coastal Vegetation and Flora of the Northern Part of the Gulf of California*. Natural History Museum of Los Angeles County Contributions in Science No. 285. Los Angeles.

Felger, Richard Stephen and Mary Beck Moser

1985 *People of the Desert and Sea: Ethnobotany of the Seri Indians*. University of Arizona Press, Tucson.

Felger, Richard S. and Benjamin T. Wilder

2009 Floristic Diversity and Long-Term Vegetation Dynamics of Isla San Pedro Nolasco, Gulf of California, Mexico. Typescript, in review for publication. (Copy in Bowen's possession).

Fladmark, K.R.

1979 Routes: Alternative Migration Corridors for Early Man in North America. *American Antiquity* 44(1):55-69.

1983 Times and Places: Environmental Correlates of Mid-to-Late Wisconsinan Human Population Expansion in North America. In *Early Man in the New World*, edited by Richard Shutler, Jr., pp. 13-41. Sage, Beverly Hills, California.

de Francia, Gonzalo

1930 [1596] Memorial of Gonzalo de Francia. In "Pearl Fishing Enterprises in the Gulf of California," by Henry R. Wagner, pp. 218-220. *Hispanic American Historical Review* 10(2):188-220.

Fujita, Harumi

1998 *Informe de la Octava Temporada de Campo del Proyecto "Identificación y Catalogación de los Sitios Arqueológicos del Área del Cabo, B.C.S."*. Archivo Técnico del Instituto Nacional de Antropología e Historia, México.

2002 *Evidencia de una Larga Tradición Cultural en la Isla Espíritu Santo, Baja California Sur*. Typescript of paper presented at the Third Binational Symposium on Balances and Perspectives: Antropología y Historia de Baja California. Mexicali. (Copy in Bowen's possession).

2006 The Cape Region. In *The Prehistory of Baja California*, edited by Don Laylander and Jerry D. Moore, pp. 82-98. University Press of Florida, Gainesville.

Fujita, Harumi (continued)
2007 En Torno a la Antigüedad de la Primera Ocupación en la Covacha Babisuri de la Isla Espíritu Santo, Baja California Sur. In *Memoria del Seminario de Arqueología del Norte de México*, edited by Cristina García M. and Elisa Villalpando C., pp. 15-24. Centro Instituto Nacional de Antropología e Historia Sonora, Hermosillo. On compact disk.

2008a *Informe Final del Proyecto "El Poblamiento de América Visto desde la Isla Espíritu Santo, B.C.S."* Archivo Técnico del Instituto Nacional de Antropología e Historia, México.

2008b *Investigaciones Arqueologicos en la Isla Cerralvo: Análisis Comparativo de Patrón de Asentamiento.* Ponencia presentada en el Segundo Encuentro de Antropología e Historia de Baja California Sur, La Paz.

Fujita, Harumi and Gema Poyatos de Paz
1998 Settlement Patterns on Espíritu Santo Island, Baja California Sur. *Pacific Coast Archaeological Society Quarterly* 34(4):67-105.

Fujita, Harumi, Miguel Téllez Duarte, and Luis Felipe Bate
2006 Una Probable Ocupación desde el Pleistoceno en la Covacha Babisuri, Isla Espíritu Santo, Baja California Sur, México. In *2° Simposio Internacional el Hombre Temprano en América*, edited by José Concepción Jiménez López, Oscar J. Polaco, Gloria Martínez Sosa, and Rocío Hernández Flores, pp. 61-72. Instituto Nacional de Antropología e Historia, México.

Gentry, Howard Scott
1949 Land Plants Collected by the *Velero III*, Allan Hancock Pacific Expeditions 1937-1941. *Allan Hancock Pacific Expeditions* 13(2):5-181. University of Southern California Press, Los Angeles.

Gerhard, Peter and Howard E. Gulick
1970 *Lower California Guidebook*, 4th ed. Arthur H. Clark, Glendale, California.

Gilg, Adam
1965 [1692] [Gilg's Letter]. In "The Seri Indians in 1692 as Described by Adamo Gilg, S.J.," translated and edited by Charles C. DiPeso and Daniel S. Matson, pp. 40-56. *Arizona and the West* 7(1):33-56.

Goebel, Ted, Michael R. Waters, and Dennis H. O'Rourke
2008 The Late Pleistocene Dispersal of Modern Humans in the Americas. *Science* 319:1497-1502.

Goss, N.S.
1888 New and Rare Birds Found Breeding on the San Pedro Martir Isle. *The Auk* 5(3):240-244.

Griffen, William B.
1959 *Notes on Seri Indian Culture, Sonora, Mexico.* Latin American Monographs 10. University of Florida Press, Gainesville.

Grismer, L. Lee
1994 Geographic Origins for the Reptiles on Islands in the Gulf of California, Mexico. *Herpetological Natural History* 2(2):17-40.

2002 *Amphibians and Reptiles of Baja California.* University of California Press, Berkeley.

Gruhn, Ruth
1988 Linguistic Evidence in Support of the Coastal Route of Earliest Entry into the New World. *Man*, New Series, 23(1):77-100.

1994 The Pacific Coast Route of Initial Entry: An Overview. In *Method and Theory for Investigating the Peopling of the Americas*, edited by Robson Bonnichsen and D. Gentry Steele, pp. 249-256. Center for the Study of the First Americans, Oregon State University, Corvallis.

Haynes, C. Vance, Jr., Paul E. Damon, and Donald C. Grey
1966 Arizona Radiocarbon Dates VI. *Radiocarbon* 8:1-21.

Hanna, G. Dallas and Leo George Hertlein
1927 *Expedition of the California Academy of Sciences to the Gulf of California in 1921: Geology and Paleontology*. Proceedings of the California Academy of Sciences, Fourth Series, 16(6):137-157.

Hardy, R.W.H.
1977 [1829] *Travels in the Interior of Mexico in 1825, 1826, 1827, & 1828*. Rio Grande Press, Glorieta, New Mexico.

Hayden, Julian D.
1956 Notes of the Archaeology of the Central Coast of Sonora, Mexico. *Kiva* 21(3-4):19-22.

Heizer, Robert F. and William C. Massey
1953 *Aboriginal Navigation off the Coasts of Upper and Baja California*. Bureau of American Ethnology Bulletin 151, Anthropological Papers 39:285-312, US Government Printing Office, Washington.

Herrera Casanova, Lorenzo
2009 Those who Had *Hast Quita* as their Birthplace. Translation and introduction by Stephen A. Marlett. In *Inside Dazzling Mountains*, edited by David Kozak. Typescript, in review for publication. (Copy in Bowen's possession).

Herrera Marcos, Roberto
1988 Yaqui Hands. In "Seri History (1904): Two Documents," introduction, notes, and translation by Mary Beck Moser, pp. 493-501. *Journal of the Southwest* 30(4):469-501.

Hills, Jim
2004 *An Introduction to Seri Indian Winds*. Paper presented at the Gulf of California Conference, June 2004, Tucson.

Hills, Jim and David Yetman
2007 A World Revealed by Language: A New Seri Dictionary and Unapologetic Speculations on Seri Indian Deep History. *Journal of the Southwest* 49(4):507-530.

Hinton, Thomas B.
1955 A Seri Girls' Puberty Ceremony at Desemboque, Sonora. *Kiva* 20(4):8-11.

Holzkamper, Frank M.
1956 Artifacts from Estero Tastiota, Sonora, Mexico. *Kiva* 21(3-4):12-19.

Hyland, Justin Robert
1997 *Image, Land, and Lineage: Hunter-Gatherer Archaeology in Central Baja California, Mexico*. Ph.D. dissertation, University of California, Berkeley.

Hyland, Justin R. and María de la Luz Gutiérrez
    1996 An Obsidian Fluted Point from Central Baja California. *Journal of California and Great Basin Anthropology* 17(1):126-128.

ten Kate, H.F.C.
    1884 Matériaux pour Servir a l'Anthropologie de la Pres-qu'ile Californienne. *Bulletin de la Société d'Anthropologie de Paris* Troisieme Série, Num. 7:551-569.

    1977 [1885] My Journey to the Peninsula of Baja California. In "In Search of the Original Californian: Herman ten Kate's Expedition to Baja California," translated and annotated in part by Peter W. van der Pas, pp. 53-81. *Journal of San Diego History* 23(3):41-92.

Kino, Francisco Eusebio
    1954 [1683] *Kino Reports to Headquarters: Correspondence of Eusebio Francisco Kino, S.J. from New Spain to Rome.* English translation and notes by Ernest J. Burrus, S.J. Institutum Historicum Societatis Jesu, Rome.

Klauber, Lawrence M.
    1949 *Some New and Revived Subspecies of Rattlesnakes.* Transactions of the San Diego Society of Natural History 11(6):61-116.

Kroeber, A.L.
    1931 *The Seri.* Southwest Museum Papers No. 6, Los Angeles.

Krutch, Joseph Wood
    1968 Isla Ildefonso. *Pacific Discovery* 21(6):18-25.

Lambeck, Kurt and John Chappell
    2001 Sea Level Change through the Last Glacial Cycle. *Science* 292:679-686.

Laylander, Don
    1992 The Development of Baja California Prehistoric Archaeology. In *Essays on the Prehistory of Maritime California*, edited by Terry L. Jones, pp. 231-250. Center for Archaeological Research at Davis Publication No. 10, Davis, California.

    2004 *Bibliography of Baja California: Prehistory -- Early History -- Ethnography.* http://pweb.jps.net/~dlaylander/recbiblio1.html.

    2006 Issues in Baja California Prehistory. In *The Prehistory of Baja California*, edited by Don Laylander and Jerry D. Moore, pp. 1-13. University Press of Florida. Gainesville.

    2007 *Databases: Archaeological Radiocarbon Dates [for Baja California].* www.pweb.jps.net/~c14dates.html.

Laylander, Don and Jerry D. Moore (editors)
    2006 *The Prehistory of Baja California.* University Press of Florida. Gainesville.

Le Boeuf, Burney J., David Aureoles, Richard Condit, Claudio Fox, Robert Gisiner, Rigoberto Romero, and Francisco Sinsel
    1983 *Size and Distribution of the California Sea Lion Population in Mexico.* Proceedings of the California Academy of Sciences 43(7):77-85.

Lewis, Leland R. and Peter E. Ebeling
   1971 *Baja Sea Guide*, Vol. II. Miller Freeman, San Francisco.

Lindsay, George E.
   1962 *The Belvedere Expedition to the Gulf of California.* Transactions of the San Diego Society of Natural History 13(1):1-44.

Lindsay, George E. and Iris H.W. Engstrand
   2002 History of Scientific Exploration in the Sea of Cortés. In *A New Island Biogeography of the Sea of Cortés*, edited by Ted J. Case, Martin L. Cody, and Exequiel Ezcurra, pp. 3-13. Oxford University Press, New York.

Lindsay, Geraldine and George Lindsay
   1981 Baja California Circumnavigated. *Pacific Discovery* 34(6):1-13.

López-Forment C., William, Irma E. Lira, and Carolina Müdespacher
   1996 *Mamíferos: Su Biodiversidad en las Islas Mexicanas.* A.G.T. Editor, México.

Lowell, Edith S.
   1970 A Comparison of Mexican and Seri Indian Versions of the Legend of Lola Casanova. *Kiva* 35(4):144-158.

Maillard, Joseph
   1923 *Expedition of the California Academy of Sciences to the Gulf of California in 1921: The Birds.* Proceedings of the California Academy of Sciences, Fourth Series, 12(24):443-456.

Mandryk, Carole A.S., Heiner Josenhans, Daryl W. Fedje, and Rolf W. Matthewes
   2001 Late Quaternary Paleoenvironments of Northwestern North America: Implications for Inland Versus Coastal Migration Routes. *Quaternary Science Reviews* 20:301-314.

Marlett, Stephen A.
   1984 Personal and Impersonal Passives in Seri. In *Studies in Relational Grammar* 2, edited by David M. Perlmutter and Carol Rosen, pp. 217-239. University of Chicago Press, Chicago.

   1990 Person and Number Inflection in Seri. *International Journal of American Linguistics* 56:503-541.

Massey, William C.
   1949 Tribes and Languages of Baja California. *Southwestern Journal of Anthropology* 5(3):272-307.

   1955 *Culture History in the Cape Region of Baja California, Mexico.* Ph.D. dissertation, University of California, Berkeley.

   1966 Archaeology and Ethnohistory of Lower California. In *Archaeological Frontiers and External Connections*, edited by Gordon F. Ekholm and Gordon R. Willey, pp. 38-58. Handbook of Middle American Indians, Vol. 4, Robert Wauchope, general editor, University of Texas Press, Austin.

Mathes, W. Michael
   1991 Francisco de Ortega's Third Voyage to the Gulf of California: A Historical Reality. *Hispanic American Historical Review* 71(1):133-135.

Mathes, W. Michael (continued)
    2006 Ethnohistoric Evidence. In *The Prehistory of Baja California*, edited by Don Laylander and Jerry D. Moore, pp. 42-66. University Press of Florida, Gainesville.

McGee, WJ
    1898 *The Seri Indians*. Seventeenth Annual Report of the Bureau of American Ethnology, Part 1. U.S. Government Printing Office, Washington.

    2000 *Trails to Tiburón: The 1894 and 1895 Field Diaries of WJ McGee*. Transcribed by Hazel McFeely Fontana, annotated and with an introduction by Bernard L. Fontana. University of Arizona Press, Tucson.

Mixco, Mauricio J,
    2006 The Indigenous Languages. In *The Prehistory of Baja California*, edited by Don Laylander and Jerry D. Moore, pp. 24-41. University Press of Florida, Gainesville.

Moser, Edward
    1963 Seri Bands. *Kiva* 28(3):14-27.

    ca. 1966 [Unpublished map of Seri camps on Isla Tiburón]. Copy in Bowen's possession.

    1973 Seri Basketry. *Kiva* 38(3-4):105-140.

Moser, Edward and Mary Beck Moser
    1961 *Vocabulario Seri: Seri-Español, Español-Seri*. Serie de Vocabularios Indigenas 5. Instituto Lingüístico de Verano, México.

    1965 Consonant-Vowel Balance in Seri (Hokan) Syllables. *Linguistics* 16:50-67.

    1976 Seri Noun Pluralization Classes. In *Hokan Studies*, edited by Margaret Langdon and Shirley Silver, pp. 285-296. Mouton, The Hague, Netherlands.

Moser, Edward and Richard S. White, Jr.
    1968 Seri Clay Figurines. *Kiva* 33(3):133-154.

Moser, Mary Beck
    1970 Seri: From Conception through Infancy. *Kiva* 35(4):201-210.

    1978 Switch Reference in Seri. *International Journal of American Linguistics* 44(2):113-120.

    1988 Seri History (1904): Two Documents. *Journal of the Southwest* 30(4):469-501.

Moser, Mary Beck and Stephen A. Marlett
    1998 How Rabbit Fooled Puma: A Seri Text. In *Studies in American Indian Languages: Description and Theory*, edited by Leanne Hinton and Pamela Munro, pp. 117-129. University of California Publications in Linguistics 131. University of California Press, Berkeley.

Moser, Mary Beck and Stephen A. Marlett (compilers)
    2005 *Comcáac quih Yaza quih Hant Ihíip hac: Diccionario Seri-Español-Inglés*. Plaza y Valdéz Editores, México.

Murphy, Robert W., Francisco Sanchez-Piñero, Gary A. Polis, and Rolf L. Aalbu
    2002 New Measurements of Area and Distance for Islands in the Sea of Cortés. In *A New Island Biogeography of the Sea of Cortés*, edited by Ted J. Case, Martin L. Cody, and Exequiel Ezcurra, pp. 447-464. Oxford University Press, New York.

Nabhan, Gary Paul

2000 Cultural Dispersal of Plants and Reptiles to the Midriff Islands of the Sea of Cortés: Integrating Indigenous Human Dispersal Agents into Island Biogeography. *Journal of the Southwest* 42(3):545-558.

2002 Cultural Dispersal of Plants and Reptiles. In *A New Island Biogeography of the Sea of Cortés*, edited by Ted J. Case, Martin L. Cody, and Exequiel Ezcurra, pp. 407-416. Oxford University Press, New York.

2003 *Singing the Turtles to Sea: The Comcáac (Seri) Art and Science of Reptiles*. University of California Press, Berkeley.

Nápoli, Ignacio María

1970 [1721] *The Cora Indians of Baja California: The Relación of Father Ignacio María Nápoli, S.J., September 20, 1721*. Translated and edited by James Robert Moriarty III and Benjamin F. Smith. Baja California Travel Series 19. Dawson's Book Shop, Los Angeles.

Nelson, Edward W.

1922 *Lower California and its Natural Resources*. Memoirs of the National Academy of Sciences, Vol. 16. U.S. Government Printing Office, Washington.

Nentvig, Juan

1980 [1764] *Rudo Ensayo: A Description of Sonora and Arizona in 1764*. Translated, clarified, and annotated by Alberto Francisco Pradeau and Robert R. Rasmussen. University of Arizona Press, Tucson.

Nolasco, Margarita

1967 *Los Seris, Desierto y Mar*. Anales del Instituto Nacional de Antropología e Historia Tomo 18. Secretaría de Educación Pública, México.

de Ortega, Francisco

1970a [1632] Descripción y Demarcación de las Islas Californias, Sondas y Catas de los Comederos de Perlas que Hay en Dichas Islas, Hecha por Mi, el Capitán Francisco de Ortega. In *Californiana II: Documentos para la Historia de la Explotación Comercial de California 1611-1679*, Vol. I, edited by W. Michael Mathes, pp. 402-424. Colección Chimalistac 29. Ediciones José Porrua Turanzas, Madrid.

1970b [1633] Segunda Demarcación de las Islas Californias Hecha por Mi el Capitán y Cabo Francisco de Ortega. In *Californiana II: Documentos para la Historia de la Explotación Comercial de California 1611-1679*, Vol. I, edited by W. Michael Mathes, pp. 425-452. Colección Chimalistac 29. Ediciones José Porrua Turanzas, Madrid.

1970c [1636] Tercera Demarcación, que Yo, el Capitán y Cabo, Francisco de Ortega Salgo a Hacer desde este Puerto de Santa Catalina, Provincia de Sinaloa, a las Islas Californias. In *Californiana II: Documentos para la Historia de la Explotación Comercial de California 1611-1679*, Vol. I, edited by W. Michael Mathes, pp. 453-467. Colección Chimalistac 29. Ediciones José Porrua Turanzas, Madrid.

Pearson, Georges A.

2004 Pan-American Paleoindian Dispersals and the Origins of Fishtail Projectile Points as Seen through the Lithic Raw-Material Reduction Strategies and Tool-Manufacturing Techniques at the Guardiría Site, Turrialba Valley, Costa Rica. In *The Settlement of the American Continents: A Multidisciplinary Approach to Human Biogeography*, edited by C. Michael Barton, Geoffrey A. Clark, David R. Yesner, and Georges A. Pearson, pp. 85-102. University of Arizona Press, Tucson.

Peltier, W.R.
  2002 On Eustatic Sea Level History: Last Glacial Maximum to Holocene. *Quaternary Science Reviews* 21:377-396.

Pérez de Ribas, Andrés
  1999 [1645] Father Andrés Pérez de Ribas on the Seris, 1645. In *Empire of Sand: The Seri Indians and the Struggle for Spanish Sonora, 1645-1803*, edited by Thomas E. Sheridan, pp. 21-23. University of Arizona Press, Tucson.

Pfefferkorn, Ignaz
  1989 [1794] *Sonora: A Description of the Province*. Translated and annotated by Theodore E. Treutlein. University of Arizona Press, Tucson.

Pimentel, Francisco Antonio
  1999 [1750] Diary of What Came to Pass and Was Executed in the Expedition that Was Made to the Island of Tiburón in this Year, 1750. In *Empire of Sand: The Seri Indians and the Struggle for Spanish Sonora, 1645-1803*, edited by Thomas E. Sheridan, pp. 178-231. University of Arizona Press, Tucson.

Porter Cassanate, Pedro [Pedro Porter y Casanate]
  1970 [1651] Carta de Pedro Porter al Virrey Conde de Alva, con Informe Adjunto. In *Californiana II: Documentos para la Historia de la Explotación Comercial de California 1611-1679*, Vol. II, edited by W. Michael Mathes, pp. 887-901. Colección Chimalistac 29. Ediciones José Porrua Turanzas, Madrid.

Reygadas Dahl, Fermín
  2003 Historia de la Arqueología en la Península de Baja California. *Arqueología Mexicana* 11(62):32-39.

Rogers, Frederick S.
  1930 [Field Notes]. Manuscript on file, San Diego Museum of Man, San Diego. (Copy in Bowen's possession).

Ryerson, Scott H.
  1976 Seri Ironwood Carving: An Economic View. In *Ethnic and Tourist Arts: Cultural Expressions from the Fourth World*, edited by N.H.H. Grayburn, pp. 119-136. University of California Press, Berkeley.

Sánchez, M. Guadalupe
  2001 A Synopsis of Paleo-Indian Archaeology in Mexico. *Kiva* 67(2):119-136.

de Salvatierra, Juan María
  1946 [1699] [Letter to Padre Juan de Ugarte, April 1, 1699]. In *Misión de la Baja California*, introduction, arrangement, and notes by Constántino Bayle, S.J., pp. 93-114. La Editorial Catolica, Madrid.

  1971 [1709] *Juan María de Salvatierra, S.J.: Selected Letters about Lower California*. Translated and annotated by Ernest J. Burrus, S.J. Baja California Travel Series 25. Dawson's Book Shop, Los Angeles.

Sheldon, Charles
  1979 [1921] *The Wilderness of Desert Bighorns & Seri Indians*. Edited and annotated by David E. Brown, Paul M. Webb, and Neil B. Carmony. Arizona Desert Bighorn Sheep Society, Phoenix.

Sheridan, Thomas E.
  1999 *Empire of Sand: The Seri Indians and the Struggle for Spanish Sonora, 1645-1803*. University of Arizona Press, Tucson.

Slevin, Joseph R.
  1923 *Expedition of the California Academy of Sciences to the Gulf of California in 1921: General Account.* Proceedings of the California Academy of Sciences, Fourth Series, 12(6):55-72.

Smith, William Neil
  1974 Seri Indians and Sea Turtles. *Journal of Arizona History* 15(2):139-158.

Spence, Tomás
  1985 [1850] Carta y Oficio de don Tomás Spence. In *Noticias Estadisticas del Estado de Sonora*, by José Francisco Velasco, pp. 147-155. Gobierno del Estado de Sonora, Hermosillo.

Stratford, Guillermo [William Strafford]
  1958 [1746] Descripción de las Californias desde el Cabo de San Lucas. In *Tres Documentos Sobre el Descubrimiento y Exploración de Baja California por Francisco María Piccolo, Juan de Ugarte y Guillermo Stratford*, edited by Roberto Ramos, pp. 52-65. Documentos para la Historia de Baja California, Número 1. Editorial Jus, México.

Taraval, Sigismundo
  1931 [1734-1737] *The Indian Uprising in Lower California, 1734-1737.* Translation, introduction, and notes by Marguerite Eyer Wilbur. The Quivera Society, Los Angeles.

Townsend, Charles Haskins
  1923 Birds Collected in Lower California. *Bulletin of the American Museum of Natural History* 48(Article 1):1-26.

de Ugarte, Juan
  1958 [1722] Relación del Descubrimiento del Golfo de California o Mar Lauretano, por el Padre Juan de Ugarte en el Año de 1722. In *Tres Documentos Sobre el Descubrimiento y Exploración de Baja California por Francisco María Piccolo, Juan de Ugarte y Guillermo Stratford*, edited by Roberto Ramos, pp. 14-50. Documentos para la Historia de Baja California, Número 1. Editorial Jus, México.

de Ulloa, Francisco
  1924 [1539] The Narrative of Ulloa. In "The Voyage of Francisco de Ulloa," by Henry W. Wagner, pp. 315-367. *California Historical Society Quarterly* 3(4):307-383.

Underwood, Jared G., Claudia J. Hernandez Camacho, David Aurioles-Gamboa, and Leah R. Gerber
  2008 Estimating Sustainable Bycatch Rates for California Sea Lion Populations in the Gulf of California. *Conservation Biology* 22(3):701-710.

Velarde, Enriqueta and Daniel W. Anderson
  1994 Conservation and Management of Seabird Islands in the Gulf of California: Setbacks and Successes. In *Seabirds on Islands: Threats, Case Studies, and Action Plans*, edited by D.N. Nettleship, J. Berger, and M. Gochfeld, pp. 229-243. Birdlife Conservation Series (1994) No. 1. International Council for Bird Preservation, Cambridge, England.

Velarde, Enriqueta, Jean-Luc E. Cartron, Hugh Drummond, Daniel W. Anderson, Fanny Rebón Gallardo, Eduardo Palacios, and Cristina Rodríguez
  2005 Nesting Seabirds of the Gulf of California's Offshore Islands: Diversity, Ecology, and Conservation. In *Biodiversity, Ecosystems, and Conservation in Northern Mexico*, edited by Jean-Luc E. Cartron, Gerardo Ceballos, and Richard Stephen Felger, pp. 452-470. Oxford University Press, New York.

Venegas, Miguel
   1979 [1757] *Noticia de la California y de su Conquista Temporal y Espiritual*, Tomo Segunda, Parte Tercera [edited by Andrés Burriel]. Obras Californianas del Padre Miguel Venegas, S.J., Vol. 2, edited by W. Michael Mathes. Universidad Autónoma de Baja California Sur, La Paz.

Villalobos Acosta, César
   2007 Proyecto Salvamento Arqueológico "Los que Viven Hacia el Verdadero Viento": Bahía Tepoca, Mar de Cortés, Sonora. In *Memoria del Seminario de Arqueología del Norte de México*, edited by Cristina García M. and Elisa Villalpando C., pp. 350-360. Centro Instituto Nacional de Antropología e Historia Sonora, Hermosillo. On compact disk.

Villalpando C., María Elisa
   1989 *Los que Viven en las Montañas*. Noroeste de Mexico 8. Centro Regional de Sonora, Instituto Nacional de Antropología e Historia, Hermosillo.

Vizcaíno, Sebastián
   1930 [1596] Vizcaíno's Narrative. In "Pearl Fishing Enterprises in the Gulf of California," by Henry R. Wagner, pp. 204-218. *Hispanic American Historical Review* 10(2):188-220.

Waters, Michael R. and Thomas W. Stafford Jr.
   2007 Refining the Age of Clovis: Implications for the Peopling of the Americas. *Science* 315:1122-1126.

White, Richard S., Jr.
   1975 [Report of a triple burial excavated on the Seri coast, March 13-15, 1975]. Manuscript on file, Centro Instituto Nacional de Antropología e Historia Sonora, Hermosillo. (Copy in Bowen's possession).

Wilder, Joseph C. (editor)
   2000 Seri Hands: A Special Issue. *Journal of the Southwest* 42(3).

Woodward, John A.
   1966 Recent Seri Culture Changes. *Masterkey* 40(1):24-32.

Xavier, Gwyneth Harrington
   1946 Seri Face Painting. *Kiva* 11(2):15-20.

www.ingramcontent.com/pod-product-compliance
Lightning Source LLC
Chambersburg PA
CBHW080405270326
41927CB00015B/3351